Current Perspectives in **Genetics**

*Insights and Applications
in Molecular, Classical,
and Human Genetics*

2000 Edition

Edited by
Shelly Cummings
University of Chicago

Brooks/Cole
Thomson Learning™

Pacific Grove • Albany • Belmont • Boston • Cincinnati • Johannesburg • London • Madrid
Melbourne • Mexico City • New York • Scottsdale • Singapore • Tokyo • Toronto

Sponsoring Editor: *Nina Horne*
Project Development Editor: *Marie Carigma-Sambilay*
Marketing Team: *Tami Cueny/Kelly Fielding*
Editorial Assistant: *John-Paul Ramin*
Production Editor: *Tessa McGlasson Avila*
Permissions Editor: *Mary Kay Hancharick*
Cover Design: *Margarite Reynolds*
Typesetting: *Diane Colwyn*
Print Buyer: *Vena Dyer*
Printing and Binding: *Globus Printing Company*

Printed in the United States of America

10   9   8   7   6   5   4   3   2   1

ISBN: 0-534-25280-X

Contents

Part Eight: Behavioral Genetics

Preface

Genetics is a relatively new specialty in the science community. However, the new and exciting discoveries that are constantly being made make genetics, especially experimental biochemical genetics, one of the most rapidly advancing fields of research. The goal in providing this reader is to help students gain a basic understanding of genetics and its varied applications in the real world. This collection of 33 recent articles from science magazines and general interest periodicals presents a wide variety of topics in genetics that influence human existence. These carefully selected articles can be used as a supplement to the material students might encounter in either human or general genetics course work. Each article begins with a brief discussion of the considerations and concerns raised by the topic. Following each article are a few questions to help students test their understanding of the material.

Science is an ongoing process so there is no guarantee that what is read as fact today will hold true tomorrow. This evolutionary process has helped researchers discover cures for many human diseases and has shed new light on a multitude of previously misunderstood behavioral conditions. I hope this reader, *Current Perspectives in Genetics - 1999 Edition - Insights and Applications in Molecular, Classical, and Human Genetics*, conveys the level of excitement that genetic research has generated in the scientific world. Moreover, after reading this collection of articles, I hope it will encourage all young scientists to share the excitement that comes from "thinking genetically."

About the Editor

Shelly Cummings is a Genetic Counselor and Senior Administrator at the University of Chicago in the Section of Hematology/Oncology. She obtained her bachelor of science degree in biology from Indiana University and her masters degree in genetic counseling from Northwestern University. She is the editor of *Current Perspectives in Genetics: Insights and Applications in Molecular, Classical, and Human Genetics* - First Edition (1995), *Current Perspectives in Genetics: Insights and Applications in Molecular, Classical, and Human Genetics* - Second Edition (1997), *Current Perspectives in Biology* - First Edition (1996), *Current Perspectives in Biology* - Second Edition (1998) and has been the author of numerous scientific publications.

Acknowledgments

The editor wishes to express her sincere thanks to the many magazines and journals that allowed me to reprint their articles. I am indebted to Diane Colwyn, who typed and produced the camera-ready pages for this book. She deserves a special thanks for waiting until the last minute before beginning production. Her round-the-clock work helped make this book as up-to-date as possible.

Shelly Cummings
University of Chicago

The Evolution of Genetics

The following time table provides a proper historical perspective for the reader concerning the relative newness of the science of genetics. Genetics, or the science of hereditary, deals with traits or characteristics that are responsible for the differences and similarities seen between individuals through generations. The late nineteenth and early twentieth centuries saw the birth of human genetics and genetic analysis. The introduction of Darwin's theory of evolution through natural selection in 1859 was the first of a series of major biological theories that marked a turning point in the history of genetics. Gregor Mendel's 1866 paper entitled "Experiments in Plant Hybridization" laid down the foundation of modern genetics. However, it was not until the 1900s that scientists began to fully understand the biological meaning of Mendel's results. The establishment of the fundamental principles of gene transmission helped expand the field of genetics. The connection between proteins and genes was first identified by the British physician Archibald Garrod in the early part of this century. Garrod was the first researcher to identify the relationship between mutant genes and malfunctioning metabolic pathways and the first person to use pedigree analysis for studying a human genetic trait.

In the early 1940s the concept of the gene as a chemical substance began taking shape with the identification of DNA as the genetic material in all organisms except certain viruses. The significant work concerning DNA that was performed by Hershey and Chase, Beadle and Tatum, Griffith, Rosalind Franklin and many others lead to the Watson and Crick hypothesis for the double-helical nature of DNA. The landmark discovery of the structure of DNA by Watson and Crick in 1953 had an immediate effect on the emerging field of molecular biology. It was established that genes controlled development. However, the genetic control elements that affect development in organisms and the details of the gene regulatory mechanisms only emerged in the 1970s, with the advent of molecular cloning techniques.

Since the development of molecular techniques, researchers have slowly begun to understand the molecular basis of human genetic diseases. The advent of somatic cell genetics as well as newer molecular methods have allowed significant advances in the understanding of human genetics. The international undertaking by the National Institutes of Health and the Department of Defense to identify the estimated 50,000 to 100,000 human genes, to locate their chromosomal position, and to sequence the entire 3 billion nucleotide pairs of DNA present in human chromosomes is one of the latest developments in genetics. Collaborative efforts are now focusing on cancer genetics and the implementation of DNA testing for men and women who may be at increased risk for developing cancer.

In September 1998, advisory committees at the Department of Energy and the National Institutes of Health approved new 5-year goals aimed at completing the Human Genome Project two years earlier than originally planned in 1990 and calls for generating a "working draft" of the human genome DNA sequence by 2001 and for obtaining the complete and highly accurate reference sequence by 2003. The target date of 2003 also will mark the 50th anniversary of Watson and Crick's description of DNA's fundamental structure. The medical industry is building upon the knowledge, resources, and technologies emanating from the Human Genome Project to further the understanding of genetic contributions to human health. As a result of this expansion of genomics into human health applications, the field of genomic medicine has been born — genetics is playing an increasingly important role in the diagnosis, monitoring, and treatment of diseases.

History of Genetics

1859	Darwin	Theory of inherited random variation and natural selection.
1865	Mendel	Laws of segregation and assortment of factors that determine different traits.
1866	Down	Clinical description of Down syndrome.
1869	Galton	Nature and nurture, Inheritance of continuously variable quantitative traits.
1872	Huntington	Clinical description of Huntington disease.
1877	Fleming	Chromosomes identified.
1882	Fleming	Human chromosome drawings.
1900	Correns, De Vries, & Tschermak	Mendel's laws rediscovered.
1900-1910	Bateson	Introduction of term "genetics." Mendel's laws applied to medicine and science.
1901	Landsteiner	ABO blood group discovered.
1901	McClung	Described role of X chromosome in human sex determination.
1901	Johannsen	Introduction of terms "gene," "genotype," and "phenotype."
1901	Punnett & Bateson	Discovery of genetic linkage.
1903	Sutton	Mendel's factors (genes) related to chromosomes.
1908	Garrod	Inborn errors of metabolism introduced as causes of disease.

1908	. . . .	Hardy & Weinberg	Extrapolated on principles of population genetics.
1910	. . . .	Rous	Description of chicken sarcoma contains disease causing agents.
1910	. . . .	Herrick	Sickle-shaped red blood cells described in anemic individuals.
1910-1920	. . . .	Morgan	Mapping of chromosomes in fruit flies.
1918	. . . .	Fisher	Additives effect of Mendelian genes described.
1920	. . . .	Fanconi	Clinical description of cystic fibrosis.
1925	. . . .	Wilson	Color blindness gene in humans assigned to X chromosome.
1926	. . . .	Vogt	Introduction of terms "expressivity" and "penetrance."
1927	. . . .	Lansteiner	MN blood group discovered.
1927	. . . .	Muller	Mutation rate increased by X-irradiation.
1928	. . . .	Griffith	Genetic alterations in pneumococci due to transformation.
1940	. . . .	Beaddle & Tatum	Introduced concept of one gene, one enzyme.
1942	. . . .	Ford	Introduction of genetic polymorphism.
1943	. . . .	Avery, McCarty & McLeod	Genetic transforming agent in pneumococci is DNA.
1949	. . . .	Barr & Bertram	Description of sex chromatin.
1949	. . . .	Pauling	Introduction of molecular diseases; sickle cell anemia.
1952	. . . .	Hershey & Chase	DNA is the hereditary material in bacteriophage.
1952	. . . .	Cori	Glucose-6-phosphatase deficient in glycogen storage disease type I.

x

1983	Waterfield, Doolittle & Duell	Oncogenes encode growth factors.
1984	Gusella	Linkage of gene for Huntington disease.
1984	Bishop	Amplified oncogenes in cancer cells.
1985	White	Described gene linkage for cystic fibrosis.
1986	Vehar & Lawn	Cloned human factor VIII gene in hamster cells.
1987-1989	Schmidt & Ledbetter	Localization of neurofibromatosis type I (NF-1) gene to chromosome 17.
1989	Blaese	Introduction of gene therapy for adenosine deaminase (ADA) deficiency.
1989	Collins & Tsui	Clones gene for cystic fibrosis.
1989		NIH Recombinant DNA Advisory Committee recommends approval of first human "gene transplant" experiment.
1990	Anderson	First attempted human gene therapy.
1990	Hall	Inherited breast cancer linked to region on long arm of chromosome 17.
1991	Hagerman & Silverman	Characterized fragile X gene
1993	Levay & Hamer	Propose a genetic influence for sexual orientation.
1993	Gusella	Localized gene for Huntington disease to chromosome 4.
1994	Miki	BRCA1 gene identified as tumor suppressor gene on chromosome 17q21.
1994		Achievement of one of the first goals of Human Genome Project: A comprehensive, high-density genetic map.

1995	Collaborative effort	Isolation of Ataxia-telangiectasis (AT) gene.
1996	Collaborative effort	Identification of susceptibility locus for prostate cancer on chromosome 1.
1997		Researchers at the National Human Genome Research Institute (NHGR) identify an altered gene that causes Pendred Syndrome. The finding may lead to a better understanding of deafness.
1998		Human Genome Project leaders announce intent to finish Human Genome sequencing two years early.
1999		Resequencing and mutational analysis using oligonucleotide microarrays.

Part One

Genetic Analysis

G.H. Hardy was an English mathematician and W. Weinberg was a German physician who first formulated the idea now known as the Hardy-Weinberg law. This theory demonstrates that genotype frequencies in a population stabilize after one generation of random mating and remain constant in future generations as long as the allele frequencies are not altered. This classical paper outlines the beginnings of this commonly used equation in genetics.

Mendelian Proportions in a Mixed Population

by G.H. Hardy

Science, July 10, 1908

To the Editor of *Science*:

I am reluctant to intrude in a discussion concerning matters of which I have no expert knowledge, and I should have expected the very simple point which I wish to make to have been familiar to biologists. However, some remarks of Mr. Udny Yule, to which Mr. R.C. Punnett has called my attention, suggest that it may still be worth making.

In the *Proceedings of the Royal Society of Medicine* (Vol. I., p. 165) Mr. Yule is reported to have suggested, as a criticism of the Mendelian position, that if brachydactyly is dominant "in the course of time one would expect, in the absence of counteracting factors, to get three brachydactylous persons to one normal."

It is not difficult to prove, however, that such an expectation would be quite groundless. Suppose that Aa is a pair of Mendelian characters, A being dominant, and that in any given generation the numbers of pure dominants (AA), heterozygotes (Aa), and pure recessives (aa) are as $p:2q:r$. Finally, suppose that the numbers are fairly large, so that the mating may be regarded as random, that the sexes are evenly distributed among the three varieties, and that all are equally fertile. A little mathematics of the multiplication-table type is enough to show that in the next generation the numbers will be as

$$(p+q)^2 : 2(p+q)(q+r) : (q+r)^2,$$

or as $p_1 : 2q_1 : r_1$, say.

The interesting question is — in what circumstances will this distribution be the same as that in the generation before? It is easy to see that the condition for this is $q^2 = pr$. And since $q_1^2 = p_1 r_1$, whatever the values of p, q and r may be, the distribution will in any case continue unchanged after the second generation.

Suppose, to take a definite instance, that A is brachydactyly, and that we start from a population of pure brachydactylous and pure normal persons, say in the ratio of $1:10,000$. Then $p=1$, $q=0$, $r=10,000$ and $p_1=1$, $q_1=10,000$, $r_1=100,000,000$. If brachydactyly is dominant, the proportion of brachydactylous persons in the second generation is $20,001:100,020,001$, or practically $2:10,000$, twice that in the first generation; and this proportion will afterwards have no tendency whatever to increase. If, on the other hand, brachydactyly were recessive, the proportion in the second generation would be $1:100,020,001$, or practically $1:100,000,000$, and this proportion would afterwards have no tendency to decrease.

In a word, there is not the slightest foundation for the idea that a dominant character should show a tendency to spread over a whole population, or that a recessive should tend to die out.

I ought perhaps to add a few words on the effect of the small deviations from the theoretical proportions which will, of course, occur in every generation. Such a distribution as $p_1:2q_1:r_1$, which satisfies the condition $q_1{}^2 = p_1 r_1$, we may call a *stable* distribution. In actual fact we shall obtain in the second generation not $p_1:2q_1:r_1$ but a slightly different distribution $p_1{}':2q_1{}':r_1{}'$, which is not "stable." This should, according to theory, give us in the third generation a "stable" distribution $p_2:2q_2:r_2$, also differing slightly from $p_1:2q_1:r_1$; and so on. The sense in which the distribution $p_1:2q_1:r_1$ is "stable" is this, that if we allow for the effect of casual deviations in any subsequent generation, we should, according to theory, obtain at the next generation a new "stable" distribution differing but slightly from the original distribution.

I have, of course, considered only the very simplest hypotheses possible. Hypotheses other than that of purely random mating will give different results, and, of course, if, as appears to be the case sometimes, the character is not independent of that of sex, or has an influence on fertility, the whole question may be greatly complicated. But such complications seem to be irrelevant to the simple issue raised by Mr. Yule's remarks.

P.S. I understand from Mr. Punnett that he has submitted the substance of what I have said above to Mr. Yule, and that the latter would accept it as a satisfactory answer to the difficulty that he raised. The "stability" of the particular ratio 1:2:1 is recognized by Professor Karl Pearson (*Phil. Trans. Roy. Soc.* (A), vol. 203, p. 60). ❑

* * * * * * * * * * * * * * * * * * * *

Questions:

1. What two factors influence the establishment of equilibrium in a randomly mating population?

2. According to Hardy's mathematical formula, assign the variables to the appropriate Mendelian characters: $p:2q:r$.

3. Why are large populations needed to full fill the expectations of the Hardy-Weinberg law?

Answers are at the end of the book.

- 2 -

The mapping of the human genome originated as a federally funded multidisciplinary endeavor involving scientists with diverse backgrounds ranging from biochemistry to anthropology. This collaborative effort, which involves some of the most advanced sequencing mechanisms, has produced a variety of maps and sequence information that must be processed in order to construct a unified linkage map. The recent announcement of a private effort to sequence the human genome four years ahead of the Human Genome Project's goal of 2005 has sparked much controversy. Some of the concerns involve whether the private effort will work, who will own the data, how quickly that data will be released, and how complete a picture of the genome that data will portray. Francis Collins, director of the National Human Genome Research Institute, has questions about whether the data generated from the two efforts will be compatible since the private effort will be using different sequencing strategies. Experts from the genetics community express their skepticism about the relationship between the private and public venture to characterize the human genome.

Privatizing the Human Genome

by Paul Smaglik

The Scientist, June 8, 1998

Principals behind joint-venture proposal and public effort seek to define relationships.

A private effort to sequence the human genome four years ahead of the Human Genome Project's 2005 goal could either compete directly with the federal project or meld seamlessly with it. Before any relationship between the two efforts becomes formalized, scientists ane federal officials involved with the Human Genome Project must determine whether the private approach will work, who will own the data, how quickly that data will be released, and how complete a picture of the genome that data will paint.

The private effort, a joint venture between Perkin-Elmer, Inc., a Norwalk, Conn., manufacturer of sequencing equipment, and J. Craig Venter, president and director of The Institute for Genomic Research (TIGR) in Rockville, Md., would be embodied by a new company formed to sequence the entire human genome by 2001, and at an estimated cost of $150 million to $300 million. Faster, more automated sequencing machines recently unveiled by Perkin-Elmer, and a different sequencing strategy espoused by Venter make that target feasible. The federally sponsored Human Genome Project, which has a total estimated budget of $3 billion, is on schedule to finish sequencing the approximately 3 billion base pairs of DNA that make up the human genome by 2005.

Francis Collins, director of the National Human Genome Research Institute, cautions that the "cheaper, better, faster" image the private project has attracted (some might say courted) may be a bit of a mirage. He notes that the federal price tag of $3 billion includes the sequencing of other genomes, necessary to do the comparative work required to ascertain each gene's function. Also, the federal project's larger budget has paid for much of the research — including gene mapping — that the private venture will use as its foundation. Collins told *The Scientist* that only about $100 million has been spent on human genome sequencing thus far. He and others are unsure whether the private product will be better — only that it will be different, since the public and private efforts will use different sequencing strategies.

That difference leads Collins to question whether the data from the two projects will be compatible. A test for whether the two sectors will indeed become "melded together," as Collins says, may be the sequencing of *Drosophila*, which the venture intends to complete as a test. Even the success of that project doesn't necessarily guarantee the success for the human approach, since the human genome dwarfs the fly's genome in size and complexity. However, failure — or the production of unusable data — from the private venture could doom further interaction. "If the strategy fails for the fly, it will almost certainly fail in the human." Finally, Collins and others concede that the method could be faster — if it works.

However, cooperation between the as-yet-unnamed company and the federal project could speed the completion of the human genome even further, Collins and others agree. Venter reports that he discussed potential collaboration scenarios with Collins, Harold Varmus, director of the National Institutes of Health, and others. However, whether that cooperation materializes remains to be seen. Reactions to the May 10 announcement about the commercial venture have been mixed.

"There are significant uncertainties about what the product will be and what its availability will be," Elbert Branscomb, director of the Joint Genome Institute in Livermore, Calif., told *The Scientist*. "What those differences are and how significant they are are not that clear."

Still, Branscomb suspects that, at minimum, the federally sponsored project will adapt to account for the private venture. The venture's potential impact — plans call for 230 automated sequencing machines to collectively churn out 100 million base pairs a day — is too potentially profound to ignore. "I think it would be surprising if the public effort did not adjust its strategy and investment." The scope and nature of that

adjustment are yet to be defined. Theoretical scenarios range from a full-blown genome sequencing race, fueled by increased federal funding, to complete cooperation between sectors, perhaps with the private effort cranking out the rough sequence, while federally funded centers fill in the gaps and do comparative work to ascertain each gene's function. The extent of adjustment and cooperation may be worked out over the next few months. All sides may have until April 1999, when the venture is scheduled to kick into full gear, to define their roles.

Data release may be one of the biggest obstacles to a full-blown collaboration between the public and private effort. Michael Hunkapiller, head of Perkin-Elmer's applied biosystems division, admits that the instrument company will, by launching this new venture, also become an information company. The new company will sell access to databases containing information about genetic variations around single genes. Pharmaceutical firms are likely to be interested in such databases. Perhaps more alarming to potential public collaborators, the company plans to also sell patent rights of genes to pharmaceutical companies and biotech firms. Tony White, Perkin-Elmer CEO, says raw data will be released to the scientific community for free. "We plan to make the genomic information widely available."

However, the frequency with which the company releases data to the public may be the key to the plan's acceptance by the scientific community. Branscomb notes that White intends a three-month lag between generating the data and releasing it for public consumption, although, "the international community has agreed to a daily release policy." That lag between discovery and release could be used to secure patents to genes before the information becomes public, Branscomb worries. For the public effort, that is a significant stumbling block.

Michael J. Morgan, program director for the Wellcome Trust of London, agrees, "We want data released every day that it comes out ..... the Sanger Centre in London, which has produced about 30 percent of the data for the Human Genome Project thus far. Days after the announcement of the private venture, the foundation doubled its support the Sanger Centre's efforts, although Morgan denies the higher commitment was a reaction to the new development. The press conference "did not leave me satisfied that this was an entirely benevolent, charitable act," Morgan says. "It is to make money."

Morgan also doubts whether Venter's proposed method, the "whole-genome shotgun approach," will work for the human genome, because the approach has only been used on much simpler organisms and results in some gaps of information. That approach does not meet with the public project's goal. "Our objective is to end up with a reference gold standard representation of the genome — largely contiguous, with few if any gaps," Morgan explains. Branscomb, too, refers to the private project's intended outcome as a "rough draft" of the complete genome. Both note that an earlier Venter proposal to try this approach was denied NIH funding in 1997 — although at the time, the new machines recently unveiled by Perkin-Elmer weren't available. Venter admits the method will result in some gaps, but emphasizes that the genome centers involved in the public project could fill those, if they so choose.

William Haseltine, CEO of Human Genome Sciences Inc., a Rockville, Md., company that formerly provided funding for TIGR, agrees that expanding the technique's application from relatively small, simple bacterial genomes to the massive and complex human genome could prove difficult. The "shotgun" technique yielded gaps in the public database for some bacteria, so applying it to humans will result in more. "There are major scale-up issues." Branscomb concurs that the proposed approach has some formidable challenges to overcome. "In doing the shotgun approach, the difficulty in getting the little bits to assemble into a whole gets worse exponentially," as the size of the genome to which it is applied increases. He also notes that the human genome has more distributed and tandem repeats than bacterial genomes, which could contribute to more gaps. But the new team has assets that could help them overcome these difficulties — assets provided by the federally funded project. The information already in the federal database — especially back clones and end sequences — will provide the scaffolding, or reference points, to hang the bits of the human genome after it is fractionated into tiny bits. "That collection of data should help the process very much."

In that respect, some collaboration already exists. Branscomb anticipates that, eventually, the federal and the private efforts will complement each other — in some as-yet-undefined way. The alternative is redundancy. "If we are truly redundant we should get out of the business and do something else." ❑

* *

Questions:

1. Why does J. Craig Venter feel his private effort to sequence the human genome will be completed before the federally funded project?

2. How many base pairs per day does the private venture plan to characterize?

3. What is the biggest disadvantage to using the "shotgun" approach to sequence the human genome?

Answers are at the end of the book.

Genetic research has advanced at a dramatic pace in the last decade — to the point where it has now become possible to attempt therapeutic genetic modification — in a few cases of human genes, where defects exist which manifests itself in certain serious diseases. This possibility, known as gene therapy, is only in its infancy. At present, no one knows how effective it will prove to be, even in the few conditions on which it is being tried — whether it will only be of relatively limited application, or whether it will open up many wider possibilities. It suffers both over optimistic claims from some quarters and exaggerated dangers from others. The suggestion of manipulating genetic changes to the human germline — such that any change was automatically passed on to all subsequent progeny — has been one which has interested many people, both over its technical and ethical aspects. A few technological optimists have speculated on eradicating certain genetic disease from affected populations, or even enhancing humanity's genetic potential. There seems to be all too many opportunities for less desirable human uses of the technology. Many people involved in ethics have raised serious doubts about the wider implications of this affecting future generations in ways in which they have no say.

One reaction has been to say that it is unlikely ever to happen. It might require dangerous and unethical experiments on human beings. Indeed, it is illegal in the United Kingdom under the Human Fertilisation and Embryology Act 1990. Other countries have established policies that forbid nontherapeutic embryo research altogether. The United States has been slow to adapt such policies, although the Recombinant DNA Advisory Committee of the National Institutes of Health will soon review its somatic cell gene therapy procedures. Overall, it is a common view that germline therapy is certainly a very distant prospect, particularly until policies are implemented to ensure that this procedure will not be used in an abusive manner.

The Politics of Germline Therapy

by Andrea L. Bonnicksen

Nature Genetics, **May 19, 1998**

Recent events signal a shift in the tenor of public discussion about human germline gene therapy (GLGT). Geneticists at a conference at UCLA outlined how GLGT might unfold over the next decade, thus lending an empirical voice to the debate on the ethics of germline genetics. Clinicians have proposed to transfer the nucleus from a patient's oocyte into an enucleated donor oocyte to circumvent a mitochondrial disorder. They have also reported the birth of an infant following the injection of ooplasm from a donor oocyte to the germ cell (unfertilized oocyte) of a patient with ooplasma deficiencies[1,2]. Both of these procedures involve nuclear transfer and cytoplasm injection which are forms of germline intervention, although it is genesplicing that is generally at the heart of ethics discussions. Finally, the Recombinant DNA Advisory Committee (RAC) of the National Institutes of Health (in the United States) is set to review its somatic-cell gene-therapy submission procedures which currently include a statement precluding the RAC from entertaining GLGT proposals. This might present an opportunity for revisiting 'germline' language

From "The Politics of Germline Therapy," by A.L. Bonnicksen,
Nature Genetics, Vol. 19, no. 1, May 1998, pp. 10-11. Reprinted with permission.

as well. As GLGT moves from hypothetical speculation to clinical feasibility, establishing forums for public discussion and debate is highly desirable.

Agreement on the ethical acceptability of germline genetics has been elusive on an international level. One group of nations, including France, Norway and Austria, restrict GLGT indirectly through policies that forbid nontherapeutic embryo research altogether. Others, such as Germany, which directly targets GLGT with its Embryo Protection Act of 1990, make germ-line interventions a criminal act[3].

A smaller number of nations have more accommodating embryo-research policies and consequently greater flexibility regarding GLGT, such as the United Kingdom, where the Human Fertilisation and Embryology Act of 1990 stipulates that licenses for embryo research will not be authorized for "altering the genetic structure of any cell" while it becomes part of an embryo *except* if it conforms with regulations (emphasis added; ref. 4). This exception, set within a framework congenial to medical innovations, leaves the door open for future germline research.

Nongovernmental organizations also offer differing positions regarding GLGT, as do two international bodies that completed documents in 1997. In one, the United Nations Educational, Social, and Cultural Organization (UNESCO) completed a Declaration on the Human Genome and Human Rights (http://www.unesco.org/ibc/uk/genome/projet/index.htm), which was signed by 186 nations. Designed to balance scientific advances with concern for human dignity, this nonbinding document calls the human genome "in a symbolic sense...the heritage of humanity," but it does not mention or proscribe GLGT. Adopting a more articulated tone, the Council of Europe has opened a Convention for the Protection of Human Rights and Dignity with Regard to Biology and Medicine (Bioethics Convention) for the signature of its 40 member states that will, with some exceptions, obligate signatories to bring their national laws into conformity with the Convention's principles. It permits preventive, diagnostic or therapeutic intervention in the human

genome only "if its aim is not to introduce any modification in the genome of descendants," which would appear to bar GLGT (ref. 5). So far, 22 nations (which do not include the United Kingdom, Germany, Russia, or Belgium) have signed this convention.

Set within this varied context, policy in the United States is unformed and minimal. Although several states have restrictive embryo research laws of untested constitutionality that would presumably forbid the experimental clinical testing of GLGT, the federal government has neither condoned nor limited human GLGT. In 1990, the Human Gene Therapy Subcommittee of the Recombinant DNA Advisory Committee (RAC) issued a "Points to Consider" document regarding somatic-cell gene research projects. One section relates to germline interventions: "The RAC and its Subcommittee will not at present entertain proposals for germline alterations...In germline alterations, a specific attempt is made to introduce genetic changes into the germ (reproductive) cells of an individual, with the aim of changing the set of genes passed on to the individual's offspring." (ref. 6).

This policy does not forbid germline alterations, but it is relevant to public funding. Notably, the policy leaves room for change by stating that such proposals will not be entertained "at present." A similar view was expressed in the 1982 report from the President's Commission for the Study of Ethical Problems in Medicine and Biomedical and Behavioral Research[7]. Acknowledging the concerns raised about interventions that would be passed to subsequent generations, the commission members nevertheless found there was no reason for "abandoning the entire enterprise." They noted that prohibitions could bring harm if they risked "depriving humanity of the great benefits genetic engineering may bring" in the future.

In light of restrictive policies worldwide, the federal government's hands-off approach, which was affirmed in 1994 when the NIH removed germline alterations from the scope of deliberations of the Human Embryo Research Panel[8], may have effectively forestalled policy questions until germline interventions are feasible. On the other hand, this approach has not encouraged protracted discussion about the conditions under which GLGT might be studied and developed. Consequently, an RAC review of Appendix M of the "Points to Consider" document, which will take place in open meetings, should invite wide public response.

Among other things, it is a good time to think about the definition of germline intervention. The definitions put forward by ethics commissions, governments and professional associations around the world range from the general (any intervention on the human germ cell) to the specific (splicing DNA sequences into nuclear DNA). Given the interest in mitochondrial disease, for example, it is unclear whether somatic-cell transfer to circumvent the inheritance of defective mitochondrial DNA falls into the same category as interventions to correct nuclear DNA. In this case, one is not altering the genetic complement of the nucleus; one is altering the match between nuclear and mitochondrial complements. The question of perspective comes into play as well — in uniting a somatic nucleus with a donor cytoplasm for the purposes of reproduction, does one transfer the nucleus (which could not receive federal funding) or does one import cytoplasm, which may or may not fall within the federal notion of germline genetics?

It is imperative that RAC deliberations on Appendix M are used as an occasion to educate policy-makers and the broader public about the science and ethics of genetic interventions. A poorly written 1997 anti-cloning law in California illustrates the danger of legislation rushed into place in the absence of full understanding of what is being restricted[9]. That law's broad definition of cloning inadvertently bans oocyte nuclear transfer, which is not cloning at all. Publicly supported discussions of GLGT may help avoid punitive and ambiguous legislative policy at the state or federal levels.

It is realistic to predict that innovations involving the human germ line will proceed with or without public inquiry and clarification. In view of this, the case for opening germline genetics to public inquiry is compelling. Currently, there is time to draw distinctions, posit rationales, and craft principles to help guide these innovations during their research and development in the private or public sectors. While some nations have foreclosed germline research, others, together with various international organizations, regard anticipatory condemnations of an entire field of research as premature and misguided. The conference at UCLA brings an important empirical reference to the discussion, and revisiting the RAC's "Points to Consider" document creates the beginnings of a policy forum for the genetics debate. ❑

References
1. Rubenstein, D.S., Thomasma, D.C., Schon, E.A. & Zinaman, M.J. *Cambridge Q. Healthcare Eth.* 4, 316-339 (1995).
2. Cohen, J., Scott, R., Schimmel, T., Levron, J. & Willadsen, S. *Lancet* **350**, 186-187 (1997).
3. German Embryo Protection Act (October 24th, 1990). Reprinted in *Hum. Repro.* **6**, 605-606 (1991).
4. Human Fertilisation and Embryology Act, 1990. Reprinted in *Int. Dig. Health Legis.* **42** 69-85 (1991).
5. Convention for the Protection of Human Rights and the Dignity of the Human Being with Regard to the Application of Biology and Medicine: Convention on Human Rights and Biomedicine. European Treaties, ETS No. 164 (1997).
6. NIH guidelines for research involving recombinant molecules (NIH Guidelines). *Fed. Reg.* **55**, 7438-7448 (1990).
7. *Splicing Life: A Report on the Social and Ethical Issues of Genetic Engineering with Human Beings.* (US Government Printing Office, Washington, DC, 1982).
8. Final Report of the Human Embryo Research Panel. (National Institutes of Health, Bethesda, Maryland, 1994).
9. Cloning of Human Beings in: *West's Annotated Californian Codes.* **40B** (1998).

* *

Questions:

1. What does the European Convention of the Protection of Human Rights and Dignity with Regard to Biology and Medicine policy states with regards to germline therapy?

2. What was the major flaw in the California anti-cloning law?

3. What would be one advantage of transferring the nucleus from one patient's oocyte into an enucleated donor oocyte?

Answers are at the end of the book.

Most people think there is only one type of DNA — the DNA that comes from the nucleus of a cell. Biologists know that DNA also exists in the mitochondria of a cell and this DNA is very unique in that it is only inherited from the mother. Forensic scientists frequently use DNA taken from crime scenes to prove innocence or guilt of a suspect. Unfortunately, the DNA left at crime scenes, such as hair, teeth, excrement, semen, and old bone are usually too badly damaged to analyze. DNA in mitochondria (mtDNA), which carry their own genes, usually remains unaltered. A specific region of the mtDNA can be sequenced to give an individual fingerprint of a person. Recently, there has been some concern about this technique and whether it is appropriate to use as evidence in criminal cases. Until this type of genetic fingerprinting is perfected, nuclear DNA analysis — which is tried and true — may hold up the best in a court of law.

Bearing False Witness

by Jonathan Knight

New Scientist, February 28, 1998

Chance matches are much more likely with mtDNA tests.

A type of genetic fingerprinting with a high chance of producing a false match has helped to convict six people in the U.S. The increasing use of this type of DNA evidence, based on DNA from the mitochondria of a cell, has sparked heated debate about whether it should be admissible in court at all.

The genetic fingerprints most people are familiar with are based on the DNA in the nucleus of the cell. Forensic scientists analyze more than a dozen different chromosomal regions, so the chances of two people having the same fingerprints are extremely low — in the order of a million to one.

But in some types of tissue found at a crime scene, such as hair without roots, teeth, excrement, and old bone, the nuclear DNA is too badly broken down to analyze. In such cases, DNA in mitochondria, which carry their own genes, often remains intact. One highly variable region of this mtDNA can be sequenced to give an individual fingerprint.

William Shields, an expert on mtDNA from the State University of New York in Syracuse, claimed last week that there are huge flaws in the way the FBI analyzes the evidence and presents it to juries. FBI expert witnesses fail to emphasise the high probability of a random match, he says. "People assume this form of DNA testing carries similar weight to other forms of DNA evidence."

Based on the FBI's own published data, however, the odds of mtDNA sequences from two people matching by chance are between 1 in 114 and 1 in 468. "I don't call that very discriminating," says Shields.

Special agent Mark Wilson of the FBI's DNA programme says that the FBI always explains these facts to juries. "We tell them straight up that this is not like nuclear DNA," he says. One problem, Wilson admits, is that the database is still rather small, containing mtDNA sequences from only a few hundred people. "But we spell it all out to the jury," he says.

Since the FBI began using the mtDNA profiles in casework in 1996, six juries in the U.S. have heard cases in which mtDNA from either the accused or the victim was presented as evidence. All six trials resulted in conviction.

A further criticism of the test is that there is now evidence that the variable mtDNA sequence used for forensic purposes can vary even in one individual. Shields says this phenomenon, called heteroplasmy, increases the chance of a random match between a suspect and a sample from the crime scene. In some cases, the mtDNA profile varies by one or two nucleotides from one end of a hair to another. "We do not yet have an adequate assessment of the frequency of heteroplasmy," he says.

But Wilson argues that, if anything, heteroplasmy is more likely to eliminate a suspect — wrongly — than incriminate an innocent person.

Because of these limitations, Britain's Forensic Science Service in Birmingham uses mtDNA only to eliminate suspects or to back up other pieces of evidence. Dave Werrett, director of research and DNA services at the FZS, says it is vital that expert witnesses explain to the jury that a random match is far more likely with mtDNA. "What we do is put all the caveats up front." ❑

* * * * * * * * * * * * * * * * * * * *

Questions:

1. What is the chance of two people having the same genetic fingerprint when DNA is taken from the nucleus of a cell?

2. Based on the FBI's own data, what are the odds of mtDNA sequences from two people matching by chance along?

3. How does Britain's Forensic Science Service use mtDNA in criminal cases?

4. What does the jury need to know about using mtDNA as evidence?

Answers are at the end of the book.

An adeno virus (AV) is a small, stable and nonpathogenic virus, which the majority of the population has been exposed to and which has never been associated with a disease. An AV is a very simple virus and contains only two genes. One is the Rep gene that codes for proteins involved in DNA replication. The other is the CAP gene, which by differential splicing, codes for three proteins that make up the protein coat of the virus. In developing a vector system, both viral genes are removed and replaced with the desired therapeutic gene; therefore, the risk of immune reaction is minimized because the vectors cannot code for its own gene products because they no longer have DNA to do so. Another advantage of AV is that it is efficient in delivering its gene to both dividing and nondividing cells. This capability is important because it allows the therapy to be used in vivo. AV vectors are produced in cells called "packaging" cells to which specific DNA "helper sequences," and the gene of interest, have been added. These helper sequences produce proteins that replicate the gene of interest and package it into a gene delivery vector with a protein coat. Vector design and applications have been frustrating to many molecular biologists due to the random uncontrolled insertion of therapeutic genetic information into unknown and unpredictable sites in the genome and lack of control of gene expression. Studies that are more recent are demonstrating that a new generation of adeno virus vectors may be on the horizon.

Have You Used an Adeno Vector...Lately?

by Alan Bernstein

Nature Genetics, April 1998

If the method for delivering new genes into cells for gene therapy were a new car, what bells and whistles would this year's new car buyers want to see? Clearly, no self-respecting tyre-kicker would choose a gas guzzler that could not deliver genes at high efficiency (and ideally only) into the necessary target cells. Proper expression of the transduced gene in the right cells, at the right time, and at appropriate levels is also important. For example, attempts to treat β-thalassemia by introduction of the β-globin gene will only work if the gene is introduced into haematopoietic stem or committed erythroid progenitor cells. Of equal importance is the degree of expression; too high a level of β-globin expression in these cells can result in α-thalassemia as a result of an imbalance in the ratio of β- to α-globin chains. Third,

because integration of exogenous DNA sequences into the host genome is mutagenic, the random uncontrolled insertion of therapeutic genetic information into unknown and unpredictable sites in the genome is an undesirable wild-card in a clinical setting. And finally, there are many diseases in which the addition of new genetic information into a cell will not, by itself, be sufficient to correct or even ameliorate the disease — this is obviously the case for dominantly-inherited diseases such as Huntington disease and most inherited cancer predisposition syndromes. Even diseases like sporadic cancers involve the somatic activation of dominantly-acting oncogenes. There are two immediate scenarios where homologous recombination (HR) would have clear advantages over nonhomologous strategies of gene

From "Have You Used an Adeno Vector...Lately?" by A. Bernstein,
Nature Genetics, Vol. 18, no. 4, April 1998, pp. 305-306. Reprinted with permission.

therapy: (i) replacement of dominant-acting mutations; and (ii) insertion of a suicide gene, placed under the strict control of a cell-type-specific regulatory region, to eliminate cancer cells by genetic ablation (ref. 1).

An approach to address many (but not all) of these challenges is familiar to mammalian and developmental biologists: the use of homologous recombination to replace specific genomic sequences with exogenous DNA. For example, it can be extremely difficult to recapitulate the expression of an endogenous cellular gene by a transgene randomly integrated into the mouse genome. Indeed, the locus-control region (LCR) involved in the complex regulation of the mammalian β-globin gene cluster was discovered during attempts to 'buffer' β-globin transgenes from neighbouring DNA sequences. In contrast, there are many examples of the accurate expression of a lacZ reporter gene dropped into a specific locus *via* homologous recombination[2]. These experiments usually involve gene transfer into a cell line or embryonic stem (ES) cells derived from the inner cell mass of a mouse blastocyst; importantly for gene-therapy considerations, adult somatic cells have also retained the capacity to carry out homologous recombination[3].

Two studies, one based on a new generation of adenovirus vectors developed by Frank Graham's group[4] and the other reported on page 325 of this issue by David Russell and Roli Hirata, who used adeno-associated virus (AAV) as a vector[5], report important advances in vector design and in the potential of combining viral vectors with HR. Russell and Hirata describe two experimental strategies to determine whether exogenous DNA, introduced into cultured human cells *via* an AAV vector, can undergo HR. To detect homologous or targeted integration events, they first infected HeLa cells harbouring a defective neomycin-resistance (*neo*) gene with an AAV vector harbouring a different insertion mutation in the same gene. HR between the two mutated genes would generate a functional gene, and hence lead to neomycin resistance. Indeed, resistant colonies were observed at a frequency of approximately 0.1% of infected cells. Analysis confirmed that these resistant colonies did indeed arise as the result of HR between the viral vector and the endogenous *neo* sequences. The frequency of *neo* gene correction was about tenfold lower than the frequency of random integration of the AAV vector, which is respectable but still below a useful HR frequency for gene therapy.

These experiments involved an artificial 'cellular' gene — the mutated *neo* gene introduced by DNA transfer prior to AAV vector infection. Thus, it could be argued that the possibility or frequency of HR cannot be extrapolated to a true single-copy cellular gene, based on these experiments. Russell and Hirata therefore infected HT-1080 fibrosarcoma cells with AAV vectors containing exons 2 and 3 of *HPRT1*, the gene that encodes hypoxanthine phosphoribo-syltransferase. Two AAV vectors were used: a control vector with a wild-type coding sequence and a mutant vector (AAV-Hpe2/3x) in which the coding region of exon 3 is disrupted by a 4–base-pair insertion. Because cells lacking HPRT are resistant to 6-thioguanine (6TG), it was possible to select HT-1080 clones that had undergone an HR event by selecting for cells that had become 6TG-resistant. At multiplicities of infection (MOI) of 50,000 vector particles per cell, 0.02–0.05% of the HT-1080 cells became 6TG-resistant as a result of HR between the AAV vector and the endogenous cellular gene. Importantly, no 6TG-resistant colonies were observed following infection with the AAV vector containing wild-type genomic sequences.

HR was also observed in normal human fibroblasts and reassuringly, the frequency of HR was dependent on the MOI. Finally, a comparison of HR frequencies of the same DNA introduced into cells by DNA transfection or AAV vector infection indicated that a much higher frequency (one thousand fold!) of HR is achieved after AAV infection. The reasons for this very large difference are not clear, nor is it established that this is a general conclusion that can be extrapolated to other genes or cells. Russell and Hirata speculate that the single-stranded nature of the AAV genome might promote homologous pairing, although other studies using single-stranded DNA substrates introduced into mammalian cells did not observe higher HR frequencise6.

This is not the first report of HR or targeted correction of cellular sequences following infection with viral vectors of mammalian cells in culture. Two previous reports demonstrated HR using adeno-virus vectors but the frequency of HR was at least 100 to 10,000 fold lower than that reported here[7,8]. Infection of cells with a retrovirus vector also resulted in targeted correction of an endogenous gene[9]. In this case, the correction appeared to involve a gene-conversion process rather than HR between viral and cellular sequences.

Targeted recombination has also been used for very different objectives. While Russell and Hirata took advantage of the host's enzymatic machinery to select for relatively rare targeted chromosomal integration events, Gudrun Schiedner and colleagues used the Cre-loxP recombination system of bacteriophage P1 to generate very high-titre adenovirus vectors in which all the viral coding sequences had been deleted and replaced by human α1-antitrypsin genomic DNA. In total, this adenovirus vector included 28 kb of human genomic DNA and no viral coding sequences. As a result, they were able to achieve high levels of tissue-specific gene expression *in vitro* and in mice over a period of more than ten months, in the absence of abnormal liver histology or hose antiviral immune response.

Together, these studies demonstrate that understanding viral life cycles and the biology of mammalian cells, coupled with the imaginative use of the recombination machinery from other species, will lead to continued marked improvements in vector design for gene therapy. Where does the field go from here? The automobile industry (and Mother Nature) has further lessons to teach us. To improve fuel efficiency and conform to more stringent antipollution standards, automobile designers have not depended on a single strategy to achieve their goals — rather, they have combined lighter materials and better engine and body design to develop cars today that are more fuel-efficient (higher titre), quieter (no host immune response), place a lighter burden on the environment (nonmutagenic) and easier to drive (appropriate gene expression) and park (targeted integration).

The burst of excitement in the mid-'80s, following the development of retrovirus vectors, was followed by a long period of frustration and disappointment. This hiatus led to the view that viral vectors may not provide the ultimate technology of choice for gene therapy. The recent spate of papers on a new generation of adenovirus vectors, and the demonstration in this issue of Nature Genetics that AAV vectors can undergo HR with single-copy chromosomal sequences, offers new hope that there is still room for further improvements in viral vector design. Will the retrovirus vector community respond to the latest challenge? ❑

References
1. Breitman, M.L. *et al. Science* **238**, 1563-1565 (1987).
2. Puri, M.C., Rossant, J., Alitalo, K., Bernstein, A. & Partanen, J. *EMBO J.* **14**, 5884-5891 (1995).
3. Brown, J.P., Wei, W. & Sedivy, J.M. *Science* **277**, 831-834 (1997).
4. Schiedner, G. *et al. Nature Genet.* **18**, 180-183 (1998).
5. Russell, D. & Hirata, R. *Nature Genet.* **18**, 323-328 (1998)
6. Fujioka, K., Aratani, Y., Kusano, K. & Koyama, H. *Nucleic Acids Res.* **21**, 407-412 (1993).
7. Wang, Q. & Taylor, M.W. *Mol. Cell. Biol.* **13**, 918-927 (1993).
8. Mitani, K. *et al. Somat. Cell Mol. Genet.* **21**, 221-231 (1995).
9. Ellis, J. & Bernstein, A. *Mol. Cell Biol.* **9**, 1621-1627 (1989).

* * * * * * * * * * * * * * * * * * * *

Questions:

1. What are two main goals that need to be met when delivering new genes into cells for gene therapy?

2. In which scenarios would homologous recombination strategies have an advantage over nonhomologous recombination strategies in gene therapy?

3. Describe how Russell and Hirata determined whether exogenous DNA, introduced into cultured human cells via an adeno-associated virus, could undergo homologous recombination.

Answers are at the end of the book.

- 6 -

Many people are asking, "What's the major goal of the Human Genome Project?" That is, why are we trying to identify all the genes in the human body and consequently develop genetic tests for those genes? At the completion of the Human Genome Project in the year 2003, researchers hope to have a better understanding of the genes that lead to human diseases and eventually develop tests to predict who may inherit these altered genes and correct these genes before they cause disease or death. Many private companies are becoming involved in the race to sequence the human genome and there are concerns that their involvement will convert the once purely scientific nature of this endeavor to a more commercial, money-making effort. The effects of any new technology introduced on the scale anticipated for biotechnology extends beyond the factories and research centers influencing our everyday lives. Biotechnology has, for example, made it possible to detect, and in some cases treat, diseases such as sickle-cell anemia, Tay-Sachs disease, cancer, and cystic fibrosis. While the economic advantages of private involvement in genetic research are obvious, the effect it may have on genetic services cannot be ignored.

The Commercialization of Human Genetics: Profits and Problems

by Timothy Caulfield

Molecular Medicine Today, April 1998

Private-sector funding is becoming increasingly important to genetic scientists and clinicians, and the number of academic-industry collaborations is growing rapidly. Furthermore, genetics has become an important tool for the healthcare industry, as the genomes of humans and other organisms are mined for new diagnostic tests and drug leads. Potentially, this is a win-win situation: academic research gets a funding boost; industry benefits from academic research; and humankind benefits from the products of these liaisons. But these benefits do not come without cost. This article explores these costs and examines whether the commercialization of academic research is compromising academic freedom, progress in clinical research, and our attitudes to normal good health.

The science of genetics has almost always had ties to industry. However, today's biomedical researchers and clinicians are experiencing more private sector involvement than ever before[1]. There has been a substantial growth in the number of academic industry collaborations[4], most of the more than 1200 patents that have been granted on human genetic sequences are privately held[5] and the biotechnology industry as a whole seems to be placing increasing emphasis on the potential of genetic research. As noted by Kevin Kinsella, CEO of Sequana Therapeutics[1], 'All biotechnology companies, by dint of the power of genomics, will have to tailor their programs toward genomics.' Indeed, the recent rise of biotechnology as a vital economic force is due, at least in part, to the excitement generated by the advances in genetic research[6,7].

Whether we are talking about the development and marketing of diagnostic services or the sale of genetic information to large pharmaceutical companies, it seems clear that human genetics has become, and will remain, a significant component of the healthcare industry. Yet, despite the benefits that can be attributed to this commercialization process, many commentators are concerned[8,9]. Will market forces adversely affect the way in which genetic services are researched, implemented and utilized? Will commercial pressures dictate the public's perception of genetics?

Amidst this environment of concern, there is another, paradoxical, reality driving commercialization. Researchers throughout the world have had to endure a curtailment in the amount of public money being devoted to both basic research and public health systems. As a result, genetic scientists are increasingly looking to the private sector to fill the ever-widening gulf between research needs and available resources. Thus there is an emerging tension. On the one hand, there are those who believe that we should slow the incursion of the market ethos into the realm of human genetics. On the other hand, private funds are becoming increasingly important to the continuation of even basic genetic research.

But what are the problems associated with commercialization? Below is a brief review of a number of the most often cited concerns. Moving from those relevant to the bench scientist (such as increased secrecy) to those with the potential to affect the clinical application of genetics (such as the premature implementation of genetic services), we see that many of these concerns are not without foundation. By understanding the nature of these issues, we can begin to formulate a balanced and thoughtful response.

The issues
Increased secrecy among genetic scientists
Though it seems clear that some form of intellectual property protection is needed to encourage investment in genetic research[6,10,11], this incentive mechanism might come with significant costs. In particular, there are those who believe the increased emphasis on obtaining patents for genetic 'inventions' has generated an atmosphere of secrecy among scientists. In order to secure the benefits of a patent, a researcher must refrain from disclosing any critical information prior to the filing of a patent application. As a result, vital scientific, or even clinically relevant[12,13], data risks being withheld for a longer period than was once the case. In fact, scientists are often required by sponsoring companies to sign 'nondisclosure' agreements. What impact will this have on the collaborative nature of scientific research?

Recent studies by Blumenthal *et al.* provide a degree of support for this concern. Their survey of US life-science academics found that 'involvement with commercialization and participation [in academic-industry research relationships] are significantly associated with the tendency to withhold the results of their research.'[14] The most-often cited reason for the delay in publication was for the purpose of filing a patent application. Interestingly, it was also found that 'investigators in the field of genetics are more likely to engage in data-withholding behaviours.'

In addition to affecting the cooperative nature of collaborative research, this increased emphasis on secrecy might also adversely impact the teaching environment. Because graduate students and research fellows are often subject to the same confidentiality agreements as principal investigators[15,16], their ability to publish in a timely fashion can be impaired.

Skewing the direction of university-based research
A related criticism of the commercialization process is that it will diminish the unique nature of university-based research. University-industry alliances, which are generally explicitly encouraged by public grant-awarding agencies, hold clear benefits for both parties; for example, they inject much-needed research funds into the universities and allow industry to benefit from university research infrastructure[17]. However, these arrangements are not without associated problems. Although the 'purity' of university research has always been under pressure from a variety of outside influences, it seems hard to deny

that the recent growth of these alliances in the realm of the life sciences has created new tensions. For example, a number of commentators have suggested that they might distort the priorities of the university away from basic research and towards that which has commercial potential[18-20]. On a more fundamental level, however, there is concern that the increased emphasis on commercialization will result in a loss of objectivity and academic freedom[21].

Commercialization might jeopardize another important function of the independent university-based researcher. Academics who do not have a financial stake in the growth of the biotechnology industry can serve as a valuable source of the expertise needed to assess the efficacy and safety of genetic technologies. If the public perceives (rightly or not) that researchers might be biased by the possibility of financial gain, the trust that is placed in academic opinions could be irreparably damaged.

Fear that genetic services will be implemented too early

There are also a variety of concerns at the level of clinical application. To date, there are only a few diagnostic services that are sold commercially. However, given the amount of financial investment in the area, the need to maintain the confidence of potential investors, and the media hype that often surrounds a new genetic discovery, many fear that the myriad of novel genetic services will be offered to the public before their efficacy has been established and the ethical, legal and social ramifications fully explored[22,23].

The commentary on the use of the *BRCA1* and *BRCA2* tests for susceptibility to breast and ovarian cancer — one of the first diagnostic services to be offered commercially — illustrates the nature of this concern. Bernadine Healy MD (Ohio State University, OH, USA), for example, has argued that 'it is too early to use *BRCA* gene testing in everyday clinical practice, because it violates a common-sense rule of medicine: don't order a test if you lack the facts to know how to interpret the results.'[24] Similar concerns have been raised in connection with tests for the *APOE4*

allele. While the presence of this gene has clearly been established as a risk factor for Alzheimer disease (AD), the test's low predictive value has led most commentators to recommend not using it as a predisposition test. However, when the gene was first identified, 'commercial interest in the United States and Europe began marketing *APOE* genotyping as a means of predicting the future development of AD in asymptomatic individuals.'[25]

The problem of premature implementation is compounded by the fact that the counseling and consent process that ought to accompany commercial services have been found to be insufficient. In a recent study evaluating the clinical use of commercial gene testing for familial adenomatous polyposis (a hereditary colon cancer syndrome caused by mutation of the *APC* gene), it was discovered that in 80% of cases the clinicians failed to provide appropriate counseling or obtain informed consent prior to testing[26].

Skewing the definition of disability, disease and normalcy

Though undoubtedly the most controversial and speculative of the dilemmas associated with the commercialization process, the idea that market forces might have a subtle eugenic effect is not without foundation. Some commentators believe that the marketing strategies that inevitably accompany commercial initiatives have the potential to affect the social perception of human normalcy[27]. For there to be a demand for a service — be it carrier testing, prenatal diagnosis, or individual testing — there must be a perceived need. What role will the commercialization process have in creating this need? Theoretically, the 'genetech' industry would benefit from a broad definition of disease and disability and a narrow view of normalcy. If commercial interests help to foster this type of definitional shift, could it be considered a 'eugenic' force? Perhaps.

It is important to note that this phenomenon does not necessarily require direct-to-consumer advertising — a marketing technique that seems likely to be tightly controlled. Indeed, the inter-industry hyping of genetic discoveries, often done in the form of press releases with the goal of attracting venture capitalists or supporting a public offering of stocks, can have a profound impact on the public's view of genetics. For example, such promotional techniques, which are frequently picked up by the popular media, arguably heighten the visibility of the 'genetic revolution' and thus validate 'the determinism on which eugenics was based.'[28] Do we want the commercial market to play such a significant, and largely unregulated, role in the development of the popular culture spin on genetics?

Although there is currently little empirical evidence to support the fear of an emerging 'laissez-faire' eugenics[29], it is hard to deny that market forces can have a profound influence on the consumption of goods. The world of healthcare is not immune to these pressures[30]. For example, the marketing of the recombinant growth hormone Protropin has arguably added to the perception that normal shortness should be considered a 'disease' that requires treatment[31,32]. Likewise, the push for prostate-specific antigen (PSA) testing has been attributed, at least to some degree, to the promotional activities of the biotechnology industry[33-35]. It is certainly possible that commercial pressures could have a similar impact on the utilization of genetic services. Whether the impact of this pressure is, in the aggregate, a 'bad thing' remains unclear. While, intuitively, one may feel skeptical of the constructive potential of market forces in this context, more research and debate seem essential.

Concluding remarks

The vibrancy and increasing importance of the biotechnology industry to many regions' economy, the pro-commercialization stance of most governments and the need for capital to undertake genetic research have created a decidedly 'anti-regulation' atmosphere. As a result, addressing the issues outlined above will undoubtedly prove to be a daunting task — one that will require a unique and creative approach. Moreover, and as already noted, many of the concerns associated with the commercialization of genetic research remain speculative. More research is certainly needed. Nevertheless, rightly or not, concerns do exist and,

at some level, they must be confronted — be it through education, regulation or otherwise. In fact, a clarification of the policy issues that surround human genetics might actually aid the commercialization process by building consumer and provider confidence. For example, it has been suggested that the unresolved nature of many of the public's legal and ethical concerns has slowed the update of *BRCA* gene testing[36].

The essential, and probably irreplaceable, role of industry should not be forgotten. There is little doubt that genetic services will ultimately be refined and disseminated through the private sector. In addition, private funding can often flow more rapidly and is more responsive and flexible than other funding mechanisms — important qualities to a field of research that moves forward as quickly as genetics. In the end, however, we must strike a balance between the legitimate needs of industry and the justifiable concerns associated with the commercialization process. ❑

References

1. Marshall, E. (1997) **The genomics gamble,** *Science* 275, 767-781
2. Miller, R. (1995) **University-industry collaboration,** *Am J. Med.* 99, 6A, 90S-92S
3. Novarro, L. (1997) **Drugs and money,** *Hospitals and Health Networks* Aug, 55-56
4. Anderson, C. (1993) **Genome project goes commercial,** *Science* 259, 300-302
5. Thomas, S.M. (1996) **Ownership of the human genome,** *Nature* 380, 387-388
6. Goudey, J. and Nath, D. (1997) **Fourth Report on the Canadian Biotechnology Industry,** Ernst and Young
7. Editorial (1996) **Capitalizing on the genome,** *Nat. Genet.* 13, 1-5
8. Kurland, C. (1997) **Beating scientists into plowshares,** *Science* 276, 761-762
9. Malinowski, M. and Blatt, R. (1997) **Commercialization of genetic testing services: the FDA, market forces and biological tarot cards,** *Tulane Law Review* 71, 1211-1312
10. Marcus, A. (1996) **Owning a gene: patent pending,** *Nat. Med.* 2, 728-729
11. Caulfield, T. *et al.* (1996) **Patent law and human DNA: current practice** in *Legal Rights and Human Genetic Material* (Knoppes, B.M., Caulfield, T. and Kinsella, T.D., eds), pp. 117-148, Edmond Montgomery
12. Percy, S. (1997) **Research conflict; doctor may be in legal battle after reporting negative findings of drug**

company study, *Medical Post* Jan, 1

13. Frankel, M. (1996) **Perception, reality, and the political context of conflict of interest in university-industry relationships,** *Academic Med.* 71, 1297-1304

14. Blumenthal, D. *et al.* (1997) **Withholding research results in academic life science: evidence from a nation survey of faculty,** *J. Am. Med. Assoc.* 277, 1224-1228

15. Blumenthal, D. *et al.* (1996) **Relationships between academic institutions and industry in the life sciences — an industry survey,** *New Engl. J. Med.* 334, 368-373

16. Blumenthal, D. (1996) **Ethics issues in academic-industry relationships in the life sciences: the continuing debate,** *Academic Med.* 71, 1291-1290

17. Haber, E. (1996) **Industry and the university,** *Nat. Biotechnol.* 14, 1501-1502

18. Blumenthal, D. (1992) **Academic-industry relationships in the life sciences,** *J. Am. Med. Assoc.* 268, 3344-3349

19. Svatos, M. (1996) **Biotechnology and the utilitarian argument for patents,** *Soc. Philos. Policy* 13, 113-144

20. Heagerty, A. (1997) **Commentary: industry-sponsored research,** *Lancet* 349, 588

21. Maatz, C. (1993) **University physician-research conflicts of interest: the inadequacy of current controls and proposed reforms,** *High Tech. Law J.* 7, 137-188

22. Milunsky, A. (1993) **Commercialization of clinical genetic laboratory services: in whose best interest?** *Obstet. Gynecol.* 81, 627-629

23. Malinowski, M. and O'Rourke, M. (1996) **A false start? The impact of federal policy on the genotechnology industry,** *Yale J. Regul.* 13, 163-254

24. Healy, H. (1997) **BRCA genes — bookmaking, fortune telling, and medical care.** *New Engl. J. Med.* 1448-1449

25. Rellkin, N., Kwon, Y., Tsai, J. and Gandy, S. (1996) **The National Institute on Aging/Alzheimer's Association recommendations on the application of apolpoprotein E genotyping to Alzheimer disease,** *Ann. New York Acad. Sci* 802, 149-171

26. Giardello, F. *et al.* (1997) **The use and interpretation of commercial APC gene testing for familial adenomatous polyposis,** *New Engl. J. Med.* 336, 823-827

27. Testart, J. (1995) **The new eugenics and medicalized reproduction,** *Cambridge Quart. of Healthcare Ethics* 4, 304-312

28. Lewontin, R.C. (1992) *Biology as Ideology: The Doctrine of DNA,* Harpers Perennial

29. Kitcher, P. (1996) *The Lives to Come: The Genetic Revolution and Humane Possibilities,* Simon and Schuster

30. Lexchin, J. (1993) **Interaction between physicians and the pharmaceutical industry: what does the literature say?** *Can. Med. Assoc. J.* 149, 1401-1407

31. Leopold, M. (1993) **The commercialization of biotechnology,** *Ann. New York Acad. Sci.* 700, 214-231

32. Daniels, N. (1997) **The Human Genome Project and the distribution of scarce medical resources,** in *The Human Genome and the Future of Healthcare* (Murray, T., Rothstein, M. and Murray, R., eds), Indiana University Press

33. Weiss, R. (1995) **Controversy surrounds marketing of PSA test,** *The Washington Post Health* 23 May, 9

34. Colburn, D. (1995) **The PSA test: too much of a good thing?** *The Washington Post Health* 23 May, 8

35. Chodak, G. (1994) **Screening for prostate cancer; the debate continues,** *J. Am. Med. Assoc.* 272, 813-814

36. Marshall, E. (1997) **Gene tests get tested,** *Science* 275, 782

* * * * * * * * * * * * * * * * * * * *

Questions:

1. What is the most often cited reason for the delay in publication of research results?

2. Who owns the greatest number of human DNA sequence patents worldwide?

3. What is Healy's basis for arguing that it is too early to offer *BRCA1* and *BRCA2* testing for the susceptibility to breast and ovarian cancer?

Answers are at the end of the book.

Part Two

The Chemistry
of
Inheritance

This classic landmark paper was the first published article delineating the double-helical form of DNA (deoxyribose nucleic acid). Watson and Crick begin this manuscript by pointing out the inconsistencies of previously proposed models for the structure of DNA. The authors describe the novel pairing and bonding between purine and pyrimidine bases. The analytical success of Watson and Crick was attributed to their ability to build a model that conformed to available data on base composition analysis of hydrolyzed samples of DNA and x-ray diffraction studies of DNA. This article greatly affected the advancing field of molecular biology and genetics. Since the publication of Watson and Crick's ideas, vast amounts of experimental research have proven their historical theory.

Molecular Structure of Nucleic Acids: A Structure for Deoxyribose Nucleic Acid

by J.D. Watson and F.H.C. Crick

Nature, 1953

We wish to suggest a structure for the salt of deoxyribose nucleic acid (D.N.A.). This structure has novel features which are of considerable biological interest. A structure for nucleic acid has already been proposed by Pauling and Corey.[1] They kindly made their manuscript available to us in advance of publication. Their model consists of three intertwined chains, with the phosphates near the fibre axis, and the bases on the outside. In our opinion, this structure is unsatisfactory for two reasons: (1) We believe that the material which gives the X-ray diagrams is the salt, not the free acid. Without the acidic hydrogen atoms it is not clear what forces would hold the structure together especially as the negatively charged phosphates near the axis will repel each other. (2) Some of the van der Waals distances appear to be too small.

Another three-chain structure has also been suggested by Fraser (in the press). In his model the phosphates are on the outside and the bases on the inside, linked together by hydrogen bonds. This structure as described is rather ill-defined, and for this reason we shall not comment on it.

We wish to put forward a radically different structure for the salt of deoxyribose nucleic acid. This structure has two helical chains each coiled round the same axis. We have made the usual chemical assumptions, namely, that each chain consists of phosphate diester groups joining β-D-deoxyribofuranose residues with 3',5' linkages. The two chains (but not their bases) are related by a dyad perpendicular to the fibre axis. Both chains follow right-handed helices, but owing to the dyad the sequences of the atoms in the two chains run in opposite directions. Each chain loosely resembles Furberg's[2] model No. 1; that is, the bases are on the inside of the helix and the phosphates on the outside. The configuration of the sugar and the atoms near it is close to Furberg's "standard configuration," the sugar being roughly perpendicular to the attached base. There is a residue on each chain every 3.4 Å in the z-direction. We have assumed an angle of 36° between adjacent residues in the same chain, so that the structure repeats after 10 residues on each chain; that is, after 34 Å. The distance of a phosphorus atom from the fibre axis is 10 Å. As the phosphates are on the outside, cations have easy access to them.

The structure is an open one, and its water content is rather high. At lower water contents we would expect the bases to tilt so that the structure could become more compact.

The novel feature of the structure is the manner in which the two chains are held together by the purine and pyrimidine bases. The planes of the bases are perpendicular to the fibre axis. They are joined together in pairs, a single base from one chain being hydrogen-bonded to a single base from the other chain, so that the two lie side by side with identical z-co-ordinates. One of the pair must be a purine and the other a pyrimidine for bonding to occur. The hydrogen bonds are made as follows: purine position 1 to pyrimidine position 1; purine position 6 to pyrimidine position 6.

If it is assumed that the basis only occur in the structure in the most plausible tautomeric forms (that is, with the keto rather than the enol configurations), it is found that only specific pairs of bases can bond together. These pairs are: adenine (purine) with thymine (pyrimidine) and guanine (purine) with cytosine (pyrimidine).

In other words, if an adenine forms one member of a pair, on either chain, then on these assumptions the other member must be thymine; similarly for guanine and cytosine. The sequence of bases on a single chain does not appear to be restricted in any way. However, if only specific pairs of bases can be formed, it follows that if the sequence of bases on one chain is given, then the sequence on the other chain is automatically determined.

It has been found experimentally[3,4] that the ratio of the amounts of adenine to thymine, and the ratio of guanine to cytosine, are always very close to unity for deoxyribose nucleic acid.

It is probably impossible to build this structure with a ribose sugar in place of the deoxyribose, as the extra oxygen atom would make too close a van der Waals contact.

The previously published X-ray data[5,6] on deoxyribose nucleic acid are insufficient for a rigorous test of our structure. So far as we can tell, it is roughly compatible with the experimental data, but it must be regarded as unproved until it has been checked against more exact results. Some of these are given in the following communications. We were not aware of the details of the results presented there when we devised our structure, which rests mainly though not entirely on published experimental data and stereochemical arguments.

It has not escaped our notice that the specific pairing we have postulated immediately suggests a possible copying mechanism for the generic material.

Full details of the structure, including the conditions assumed in building it, together with a set of co-ordinates for the atoms, will be published elsewhere.

We are much indebted to Dr. Jerry Donohue for constant advice and criticism, especially on interatomic distances. We have also been stimulated by a knowledge of the general nature of the unpublished experimental results and ideas of Dr. M.H.F. Wilkins, Dr. R.E. Franklin and their coworkers at King's College, London. One of us (J.D.W.) has been aided by a fellowship from the National Foundation for Infantile Paralysis. ❑

References

[1]Pauling, L., and Corey, R.B., *Nature*, **171,** 346 (1953); *Proc. U.S. Nat. Acad. Sci.*, 39, 84 (1953).

[2]Furberg, S., *Acta Chem. Scand.*, **6,** 634 (1952).

[3]Chargaff, E., for references see Zamenhof, S., Brawerman, G., and Chargaff, E., *Biochim. et Biophys. Acta*, **9,** 402 (1952).

[4]Wyatt, G.R., *J. Gen. Physiol*, **36,** 201 (1952).

[5]Astbury, W.T., Symp. Soc. Exp. Biol. 1, Nucleic Acid, **66** (Camb. Univ. Press, 1947).

[6]Wilkins, M.H.F., and Randall, J.T., *Biochim. et Biophys. Acta*, **10,** 192 (1953).

* * * * * * * * * * * * * * * * * * * *

Questions:

1. What are the fundamental differences between the structural model proposed by Pauling and Corey and the model suggested by Watson and Crick?

2. How are the two chains of the helical structure held together?

3. Why would the Watson and Crick model be impossible to build with a ribose sugar?

Answers are at the end of the book.

- 8 -

Over half of all tumor cells contain tiny extra chromosomes that encourage the cancer to grow. Called double minute chromosomes, the renegade DNA contains genes that have been duplicated and subsequently broken away from the chromosome where they are usually found. Double minute chromosomes have only been seen in cancer cells. Without double minute chromosomes, some researchers believe that many tumors would stop growing. Now scientists may be able to find a way to rid cells of the excess DNA, using a technique that lets them watch these chromosomes at work in living cells. This report describes how researchers took advantage of a set of proteins called histones found in all eukaryotic cells. The group fused a human histone gene to the gene for the protein that gives certain jellyfish their greenish glow. This made all the chromosomes of the cell cultures, including the double minutes, glow bright green. During cell division the researchers saw double minutes gather together and latch onto the normal chromosomes. By hitching this way, the double minutes get passed on to all a cancer cell's progeny. This technique could be used to screen for drugs that eliminate double minutes.

Histone-GFP Fusion Protein Enables Sensitive Analysis of Chromosome Dynamics in Living Mammalian Cells

by Teru Kanda, Kevin F. Sullivan, and Geoffrey M. Wahl

Current Biology, 1998

Background

In eukaryotes, segregation of sister chromatids during mitosis requires spindle fiber attachment to the kinetochores formed at centromeric DNA. Cancer cells often harbor abnormal chromosomes such as dicentric or acentric chromosomes, uneven dist. however, which would be expected to behave among daughter cells anomalously. Double minute chromosomes (Dms) are paired chromatin bodies that have been reported in as many as 50% of human tumors but have never been observed in normal cells [1, 2]. As DMs lack functional centromeres, they do not segregate by the same mechanism used by normal chromosomes. DMs can accumulate to a high copy number because of their autonomous replication during the DNA synthesis (S) phase of the cell cycle and their subsequent uneven distribution to daughter cells during mitosis. As DMs contain a diversity of amplified oncogenes [3], their uneven segregation and accumulation increases the malignant potential during tumor progression. On the other hand, as DM loss can decrease tumor cell viability [4], understanding the mechanism of DM segregation could lead to the identification of highly selective anti-neoplastic agents that specifically disrupt the transmission of DMs to daughter cells.

Observations of fixed chromosomes in DM-harboring cancer cells have provided some insights into DM segregation during mitosis [5, 6]. Fixation and permeabilization of cells may cause artificial distortions of chromosome distribution, however, and could perturb intracellular structures. An ideal strategy for examining the dynamics of DM segregation would involve their direct visualization in living cells. DMs vary in size, however, and many are at the size limit of conventional cytogenetics ($\sim 1-2 \times 10^6$ bp) [2], which has prevented their detection in cycling cells. This report describes a fluorescent labeling system with sufficient sensitivity to visualize DMs *in vivo*, and that enables analyses of their segregation dynamics in real time during mitosis. One approach to label chromosomes in living cells involves the fluorescent tagging of proteins that localize to chromosomes. The nucleosome is the fundamental repeating unit of chromatin. Each nucleosome core particle consists of an octamer of core histones with 146 bp of micrococcal-nuclease-resistant DNA wrapped around it [7]. As histones are the principal structural proteins of eukaryotic chromosomes, they are attractive targets for fluorescent labeling. Purified calf thymus histones (H2A and H2B) conjugated with rhodamine have been microinjected into *Drosophila* embryos to analyze cell lineage relationships [8] and chromosomal condensation and decondensation events [9]. The success of this approach demonstrates the utility of fluorescently labeled histones to study chromosomal dynamics in living cells.

The green fluorescent protein (GFP) of the jellyfish *Aequorea victoria* retains its fluorescent properties when recombinant GFP proteins are expressed in eukaryotic cells [10]. GFP fusion proteins have been successfully targeted to specific subcellular organelles and structures including the nucleus, plasma membrane, mitochondrion, cytoskeleton, and Golgi apparatus [11–13]. Recently, GFP tagging also enabled visualization of specific chromosomal regions [14–16]. These results indicate the potential utility of a histone–GFP fusion protein to fluorescently label chromosomes in living cells. The feasibility of this approach is indicated by the observation that a fusion protein of GFP and yeast histone H2B localized properly in yeast nuclei [17]. Here, we show that a fusion protein of GFP and human H2B (H2B–GFP) is incorporated into Nucleosome core particles without perturbing cell cycle progression. H2B–GFP bound chromosomes and DMs with

high specificity, allowing them to be easily observed using a confocal microscope. We describe for the first time the behavior of DMs during mitosis in living cells. Our results reveal that DMs often cluster in anaphase cells and attach to groups of segregating chromosomes. Sometimes, segregating daughter chromosomes are connected by DM 'bridges' spanning the midplane of anaphase cells. Time-lapse observation revealed that cytokinesis severs the DM bridges, resulting in asymmetric distribution of DMs to the daughter cells.

Results
Stable expression of H2B–GFP in HeLa cells

The cDNA encoding human H2B was tagged at its carboxyl terminus with DNA encoding codon-optimized enhanced GFP [18], and the chimeric gene was subcloned into a mammalian expression vector. The construct was introduced into the human HeLa cell line by transient transfection, and fluorescence microscopic observation indicated that H2B–GFP protein localized to interphase nuclei and mitotic chromosomes (data not shown). To analyze the effects of constitutive H2B-GFP expression on cell cycle progression, we transfected HeLa cells and cultured them under drug selection (blasticidin) to obtain clones that stably expressed the H2B–GFP transgene. GFP-positive colonies arose in about 10% of blasticidin-resistant colonies, while other colonies (~90%) were negative for GFP for unknown reasons. We obtained several stable cell lines expressing H2B–GFP. A cell line with uniform, high-level expression of H2B–GFP was chosen for further analyses. The expression level of H2B–GFP in this cell line was stable for more than three months in the absence of continuous blasticidin selection. The high degree of stability of the integrated H2B–GFP gene in the absence of selection strongly suggests that chromosome stability is unimpaired by constitutive H2B–GFP expression. The mitotic index and the growth rate of this cell line were similar to those of the parental HeLa cells (data not shown). We also used a cell line stably expressing a fusion protein of H2B fused at its amino terminus to GFP and got essentially the same results (data not shown).

H2B–GFP is incorporated into nucleosomes

We biochemically fractionated nucleosome core particles from cells expressing H2B–GFP and analyzed for the presence of the fusion protein to determine if it was a component of nucleosome core particles. Mononucleosomes were generated by extensive micrococcal nuclease digestion of the isolated nuclei expressing H2B–GFP. Digested chromatin was pelleted by subsequent centrifugation of the nuclease-treated nuclei. The supernatant and pellet, together with whole cell lysate of HeLa cells and HeLa cells expressing H2B–GFP, were analyzed by western blotting using an anti-human-H2B antibody. The majority of the expressed H2B–GFP protein was recovered in the pelleted chromatin fraction, and little if any was detected in the supernatant. The chromatin fraction was further fractionated by sucrose gradient centrifugation in the presence of 0.5 M NaCl to dissociate histone H1 [19]. Electrophoretic analysis of the DNA showed that fractions 2 through 4 (predominantly fraction 3) contained DNA of about 146 bp (the size expected for the DNA wrapped around nucleosome core particles. The proteins in the samples were analyzed by gel electrophoresis; H2B–GFP protein and core histones were identified in the mononucleosome fractions by Coomassie staining. Aliquots of the same samples were analyzed by western blotting using anti-human-H2B antibody to specifically detect H2B and H2B–GFP protein. The results demonstrate that the H2B–GFP fusion protein is in the mononucleosome fractions of the sucrose gradient and that its distribution parallels that of native histones in the gradient. The relative amounts of H2B–GFP and endogenous H2B in the purified mononucleosomes were comparable to their relative amounts in the whole cell lysate, suggesting that H2B–GFP protein is incorporated into nucleosomes efficiently. H2B–GFP association with the nucleosome core particle was stable under conditions that dissociate histone H1 [19], suggesting that adventitious aggregation of H2B–GFP protein with chromatin is unlikely.

H2B–GFP incorporation does not inhibit cell cycle progression

It was conceivable that GFP tagging of H2B protein could affect chromatin structure and perturb cell cycle progression as a consequence. Therefore, the cell cycle distribution of the established cell line expressing H2B–GFP was analyzed to ascertain differences in cell cycle progression relative to the parental cell population. Asynchronous HeLa cells and the transformant expressing H2B–GFP were fixed with ethanol, stained with propidium iodide (PI), and analyzed by fluorescence-activated cell sorting (FACS). The green emission of GFP-labeled cells produced an approximately three-log shift from parental HeLa cells. DNA content was determined by measuring the red emission of PI. The results indicate that the cell cycle distribution of asynchronous HeLa cells expressing H2B–GFP is indistinguishable from that of the parental HeLa Cells, clearly demonstrating that the H2B–GFP protein has little, if any, effect on cell cycle progression.

H2B–GFP decorates chromosomes in living cells

Cells expressing H2B–GFP were observed using confocal microscopy to determine the pattern of chromatin staining in interphase and mitosis. H2B–GFP enabled highly sensitive chromatin detection in all phases of the cell cycle. Fixation and permeabilization of the cells, which might cause artificial distortion of intracellular structure, was not required to obtain such images. H2B–GFP was highly specific for nuclear chromatin as no fluorescence was observed in the cytoplasm. In addition, H2B–GFP provided a remarkable level of sensitivity. For example, a chromatin structure that appeared to e a pair of lagging sister chromatids with a centromeric constriction was readily observed. The fine intranuclear chromatin architecture in interphase nuclei visualized by H2B–GFP was consistent with the previously reported deconvoluted optical-sectioning images of fixed nuclei obtained by 4',6'-diamidino-2-phenylindole (DAPI) staining [20]. Chromosome spreads of the H2B–GFP-expressing cells also showed that the GFP fluorescence patterns were identical to the patterns obtained using DAPI.

We also observed perinucleolar regions densely stained with H2B–GFP that resembled chromocenters in interphase nuclei. A previously described feature of chromocenters is that they are heterochromatic and often contain centromeres [21, 22]. Double staining with centromere antibodies and H2B–GFP demonstrated that certain regions with intense H2B–GFP staining possessed multiple centromeres. From this result, coupled with the concordance of H2B–GFP staining and DAPI staining, we conclude that H2B–GFP staining reflects the density of packing of DNA in different regions of the nucleus. Thus, chromosomal domains that have previously only been examined in fixed cells may be monitored using the H2B–GFP method in living cells.

Visualization of DMs in living cells

We applied our highly sensitive H2B–GFP chromatin-labeling technique to the analysis of DMs in living cells. A retroviral vector was constructed to enable the efficient transfer and expression of H2B–GFP in a broad range of host cells. We used a vesicular stomatitis virus G glycoprotein (VSV-G) pseudotyped retroviral vector to obtain high viral titers [23]. COLO320DM cells harboring DMs containing an amplified c-myc gene [24, 25] were infected with the H2B–GFP retrovirus, and two days later over 90% of the cells expressed H2B–GFP protein. FACS analyses revealed that cell cycle progression of COLO320DM cells was not affected by H2B–GFP expression (data not shown). We collected serial-sectioning images of living COLO320DM cells expressing H2B–GFP using confocal microscopy. We noticed that small fluorescent dots were frequently observed in mitotic cells. The sizes of these dots were 0.7–0.85 μm in diameter, corresponding to the size of DMs in this cell line. We found that they frequently associated in clusters in anaphase cells where they were attached to normal chromosomes. Sometimes they aligned in regular arrays and formed an extended bridge between segregating groups of daughter chromosomes.

In order to confirm that these dot-like chromatin bodies were DMs, mitotic COLO320DM cells were fixed with or without colcemid treatment, and the DM distribution was analyzed by fluorescence *in situ* hybridization (FISH) using a c-myc cosmid probe. Whereas chromosome spreads of colcemid-treated cells usually had dispersed DMs, DMs were observed to form clusters in untreated mitotic cells. The DMs detected using FISH were strikingly similar to the dot-like structures observed in H2B–GFP-expressing cells. The results strongly suggest that the dot-like chromatin bodies observed in living cells are DMs. We found that most of the observed mitotic cells contained clustered DMs, although the number of DMs varied from cell to cell. Approximately 30% of the anaphase cells showed bridge formations involving DMs. We therefore conclude that the clustering of DMs and their unbalanced distribution to daughter cells during mitosis are very common events in this cell line.

We next analyzed DM segregation by making time-lapse observations using an epifluorescence microscope. Clustered DMs were attached by the extended arms of normal segregating chromosomes, forming a chromosome bridge. This bridge was further extended as daughter chromosomes segregated. Subsequently, the bridge was severed by the process of cytokinesis, and the cluster of DMs appeared to be unevenly distributed to daughter nuclei. These results clearly demonstrate that DMs frequently cluster in anaphase cells, sometimes forming chromosomal bridges, and that their uneven distribution to daughter cells can result from cytokinesis severing DM bridges asymmetrically.

Discussion

We have described a novel strategy to fluorescently label chromosomes in living cells and the successful application of this strategy to observe DMs in living cells. Despite the large size of the GFP tag (239 amino acids), it has been shown in numerous cases that GFP-tagged proteins are functional and localize properly [11–13]. Such is also the case with the histone H2B–GFP fusion protein. Our experimental data demonstrate the co-fractionation of H2B–GFP with mononucleosomes under high ionic strength (0.5 M NaCl). The primary interactions responsible for the stability of nucleosomes are electrostatic as nucleosomes can be dissociated into their DNA and histone components by elevating ionic strength. Histone H1, as well as non-histone proteins, dissociates from the nucleosomes at 0.5 M NaCl [7, 19]. It is therefore likely that H2B–GFP protein is incorporated into nucleosome core particles rather than just affiliating with the core particles by a non-specific electrostatic interaction.

The strategy described in this paper offers significant advantages over other chromosome-labeling methods. Although fluorescent labeling of mammalian chromosomes in living cells has been demonstrated using Hoechst 33342 [26, 27], each cell line must be analyzed individually to optimize the time of drug exposure and concentration of the drug [28]. Furthermore, as Hoechst 33342 is excited maximally near 350 nm and high intensities of ultraviolet (UV) irradiation can damage cells and produce cell cycle delay or arrest, the level of UV excitation must be controlled carefully. In addition, Hoechst 33342 affects cell cycle progression, arresting cells in G2 phase [29]. Intercalating DNA drugs, like dihydroethidium, may cause mutations in the DNA by interfering with DNA replication [28]. Microinjection of rhodamine-labeled histones, successfully used in *Drosophila* [8, 9], is not suitable for analyzing a large population of mammalian cells. In contrast to these methods, the enhanced GFP [18] used in this study is excited with blue light (490 nm), which is less damaging than the UV-light excitation required for Hoechst. Moreover, constitutive expression of H2B–GFP from the integrated transgene enables long-term analyses without perturbing cell cycle progression. As the primary structures of histone proteins are well conserved among different species [30], it is likely that the H2B–GFP described here will be useful for cells of different species. This notion is supported by our ability to use the H2B–GFP vector in mouse, hamster, and monkey cells (data not shown).

The H2B–GFP strategy described in this report has a wide variety of applications for studying chromosome dynamics. For example, it can be used as a transfection marker that readily enables one to identify mitotic cells using fluorescence microscopy. H2B–GFP fluorescence persists in cells fixed in ethanol, which is useful for FACS analyses. As the intensity of H2B–GFP fluorescence depends on the chromosome condensation states in interphase cells, one can study chromosome condensation and decondensation in live cells [9]. The method may be especially useful for real-time analysis of apoptosis by enabling visualization of chromatin fragmentation and hypercondensation in living cells, as well as for studies of the effects of oncogenes on chromosome stability during tumor progression [31].

Using a retroviral vector expressing H2B–GFP, we have demonstrated that both normal chromosomes and small DMs can be readily distinguished in live cells. H2B–GFP enabled the observation of chromosomes in their native state without the need for fixation and permeabilization procedures, which can cause artificial distortion of intracellular structures. Our observations validate and extend a previous report in which DMs in fixed mitotic cells were observed to associate with the condensed chromosomes [5, 6]. This led to a model for DM segregation involving 'hitch-hiking' of DMs on chromosomes [6]. The distinctive clustering behavior of DMs and their close association with normal chromosomes provide one mechanism by which DMs are transmitted to daughter cells even though they lack functional centromeres. Clustered DMs appear to transmit to daughter cells in a highly stochastic manner, however. As COLO320DM cells grow faster when their levels of c-*myc* expression are elevated [4], daughter cells containing more DMs should be selected, and over time the population of cells will come to contain many DMs per cell. Therefore, although the number of DMs per cell is continuously changing by stochastic uneven distribution to daughter cells, DMs appear to be stably maintained when one considers the entire population. Sometimes, clustered DMs associate with both groups of the segregating daughter chromosomes, forming bridges across the midplane of anaphase cells. It is tempting to speculate that such chromosomal bridges could increase chromosomal instability by inducing non-disjunction or, in some cases, by preventing chromosome segregation and increasing the probability of a chromosome arm being severed during cytokinesis. Were breakage to occur within an arm, genomic instability could be further increased by inheritance of the broken chromosome, which could initiate a bridge-breakage fusion cycle [32].

The mechanism of DM clustering during mitosis remains to be elucidated. One possibility is that DMs cluster due to their interaction with the spindle microtubles, as interaction of non-centromeric chromatin with the mitotic spindle is well documented [33]. Chromosome spreads of colcemid-treated cells reveal a scattered distribution of DMs, suggesting that spindle microtubles may play an important role in DM clustering. Alternatively, DMs themselves may have cohesive properties and may 'stick' to each other. It has been reported that a hamster cell line containing large tandemly repeated amplicons including the dihydrofolate reductase gene also had anaphase-bridge formations due to delayed sister-chromatid disjunction [34]. Repeated arrays of amplicons in DMs may have configurations that favor DM clustering and delayed disjunction of sister minute chromosomes during anaphase. The interaction between DMs and normal chromosomes must be clarified as well. The H2B–GFP system described here should facilitate the additional analyses required to understand the precise mechanism of DM clustering and segregation.

Conclusions

We have established a novel system for labeling chromatin in living cells using a fusion protein of histone H2B and GFP. The H2B–GFP system allows chromosomes, including DMs, to be imaged at a high resolution without perturbing cell cycle control or intracellular structures. The application of this system has revealed the distinctive clustering behavior of DMs in living mitotic cells. We propose that DM clustering is an important factor leading to their asymmetric distribution to daughter cells. ❏

References

1. Benner SE, Wahl GM, Von Hoff DD: **Double minute chromosomes and homogeneously staining regions in tumors taken directly from patients versus in human tumor cell lines.** *Anticancer Drugs* 1991, **2**:11-25.
2. Cowell JK: **Double minutes and homogeneously staining regions: gene amplification in mammalian cells.** *Annu Rev Genet* 1982, **16**:21-59.
3. Stark GR, Wahl GM: **Gene amplification.** *Annu Rev Biochem* 1984, **53**:447-491.
4. Von Hoff DD, McGill JR, Forseth BJ, Davidson KK, Bradley TP, Van DD, *et al.*: **Elimination of extrachromosomally amplified MYC genes from human tumor cells reduces their tumorigenicity.** *Proc Natl Acad Sci USA* 1992, **89**:8165-8169.
5. Barker PE, Hsu TC: **Are double minutes chromosomes?** *Exp Cell Res* 1978, **113**:456-458.
6. Lavan A, Levan G: **Have double minutes functioning centromeres?** *Hereditas* 1978, **88**:81-92.
7. Wolffe A: *Chromatin: Structure and Function*, 2nd edn. San Diego: Academic Press; 1995.
8. Minden JS, Agard DA, Sedat JW, Alberts BM: **Direct cell lineage analysis in *Drosophila* melanogaster by time-lapse, three-dimensional optical microscopy of living embryos.** *J Cell Biol* 1989, **109**:505-516.
9. Hiraoka Y, Minden JS, Swedlow JR, Sedat JW, Agard DA: **Focal points for chromosome condensation and decondensation revealed by three-dimensional *in vivo* time-lapse microscopy.** *Nature* 1989, **342**:293-296.
10. Chalfie M, Tu Y, Euskirchen G, Ward WW, Prasher DC: **Green fluorescent protein as a marker for gene expression.** *Science* 1994, **263**:802-805.
11. Cubitt AB, Heim R, Adams SR, Boyd AE, Gross LA, Tsien RY: **Understanding, improving and using green fluorescent proteins.** *Trends Biochem Sci* 1995, **20**:448-455.
12. Gerdes HH, Kaether C: **Green fluorescent protein: applications in cell biology.** *FEBS Lett* 1996, **389**:44-47.
13. Misteli T, Spector DL: **Applications of the green fluorescent protein in cell biology and biotechnology.** *Nat Biotechnol* 1997, **15**:961-964.
14. Robinett CC, Straight A, Li G, Willhelm C, Sudlow G, Murray A, *et al.*: ***In vivo* localization of DNA sequences and visualization of large-scale chromatin organization using lac operator/repressor recognition.** *J Cell Biol* 1996, **135**:1685-1700.
15. Straight AF, Belmont AS, Robinett CC, Murray AW: **GFP tagging of budding yeast chromosomes reveals that protein-protein interactions can mediate sister chromatid cohesion.** *Curr Biol* 1996, **6**:1599-1608.
16. Shelby RD, Hahn KM, Sullivan KF: **Dynamic elastic behavior of alpha-satellite DNA domains visualized *in situ* in living human cells.** *J Cell Biol* 1996,

135:545-557.

17. Flach J, Bossie M, Vogel J, Corbett A, Jinks T, Willins DA, et al.: **A yeast RNA-binding protein shuttles between the nucleus and the cytoplasm.** *Mol Cell Biol* 1994, **14**:8399-8407.

18. Yang TT, Cheng L, Kain SR: **Optimized codon usage and chromophore mutations provide enhanced sensitivity with the green fluorescent protein.** *Nucleic Acids Res* 1996, **24**:4592-4593.

19. Laybourn PJ, Kadonaga JT: **Role of nucleosomal cores and histone H1 in regulation of transcription by RNA polymerase II.** *Science* 1991, **254**:238-245.

20. Belmont AS, Bruce K: **Visualization of G1 chromosomes: a folded, twisted, supercoiled chromonema model of interphase chromatid structure.** *J Cell Biol* 1994, **127**:287-302.

21. Moroi Y, Hartman AL, Nakane PK, Tan EM: **Distribution of kinetochore (centromere) antigen in mammalian cell nuclei.** *J Cell Biol* 1981, **90**:254-259.

22. Bartholdi MF: **Nuclear distribution of centromeres during the cell cycle of human diploid fibroblasts.** *J Cell Sci* 1991, **99**:255-263.

23. Burns JC, Friedmann T, Driever W, Burrascano M, Yee JK: **Vesicular stomatitis virus G glycoprotein pseudotypes retroviral vectors: concentration to very high titer and efficient gene transfer into mammalian and nonmammalian cells.** *Proc Natl Acad Sci USA* 1993, **90**:8033-8037.

24. Alitalo K, Schwab M, Lin CC, Varmus HE, Bishop JM: **Homogeneously staining chromosomal regions contain amplified copies of an abundantly expressed cellular oncogene (c-myc) in malignant neuroendocrine cells from a human colon carcinoma.** *Proc Natl Acad Sci USA* 1983, **80**:1707-1711.

25. Shimizu N, Kanda T, Wahl GM: **Selective capture of acentric fragments by micronuclei provides a rapid method for purifying extrachromosomally amplified DNA.** *Nat Genet* 1996, **12**:65-71.

26. Belmont AS, Braunfeld MB, Sedat JW, Agard DA: **Large-scale chromatin structural domains within mitotic and interphase chromosomes *in vivo* and *in vitro*.** *Chromosoma* 1989, **98**:129-143.

27. Hiraoka Y, Haraguchi T: **Fluorescence imaging of mammalian living cells.** *Chromosome Res* 1996, **4**:173-176.

28. Arndt-Jovin DJ, Jovin TM: **Fluorescent labeling and microscopy of DNA.** *Methods Cell Biol* 1989, **30**:417-448.

29. Tobey RA, Oishi N, Crissman HA: **Cell cycle synchronization: reversible induction of G2 synchrony in cultured rodent and human diploid fibroblasts.** *Proc Natl Acad Sci USA* 1990, **87**:5104-5108.

30. Wells DE: **Compilation analysis of histones and histone genes.** *Nucleic Acids Res* 1986, **14**:R119-R149.

31. Denko N, Stringer J, Wani M, Stambrook P: **Mitotic and post-mitotic consequences of genomic instability induced by oncogenic Ha-ras.** *Somat Cell Mol Genet* 1995, **21**:241-253.

32. Coquell A, Pipiras E, Toledo F, Buttin G, Debatisse M: **Expression of fragile sites triggers intrachromosomal mammalian gene amplification and sets boundaries to early amplicons.** *Cell* 1997, **89**:215-225.

33. Fuller MT: **Riding the polar winds: chromosomes motor down east.** *Cell* 1995, **81**:5-8.

34. Warburton PE, Cooke HJ: **Hamster chromosomes containing amplified human alpha-satellite DNA show delayed sister chromatid separation in the absence of *de novo* kinetochore formation.** *Chromosoma* 1997, **106**:149-159.

35. Zhong R, Roeder RG, Heintz N: **The primary structure and expression of four cloned human histone genes.** *Nucleic Acids Res* 1983, **11**:7409-7425.

36. Mizushima S, Nagata S: **pEF-BOS, a powerful mammalian expression vector.** *Nucleic Acids Res* 1990, **18**:5322.

37. Izumi M, Miyazawa H, Kamakura T, Yamaguchi I, Endo T, Hanaoka F: **Blasticidin S-resistance gene (bsr): a novel selectable marker for mammalian cells.** *Exp Cell Res* 1991, **197**:229-233.

38. Naviaux RK, Costanzi E, Haas M, Verma IM: **The pCL vector system: rapid production of helper-free, high-titer, recombinant retroviruses.** *J Virol* 1996, **70**:5701-5705.

39. Chen C, Okayama H: **High-efficiency transformation of mammalian cells by plasmid DNA.** *Mol Cell Biol* 1987, **7**:2745-2752.

40. Miyoshi H, Takahashi M, Gage FH, Verma IM: **Stable and efficient gene transfer into the retina using an HIV-based lentiviral vector.** *Proc Natl Acad Sci USA* 1997, **94**:10319-10323.

41. Sullivan KF, Hechenberger M, Masri K: **Human CENP-A contains a histone H3 related histone fold domain that is required for targeting to the centromere.** *J Cell Biol* 1994, **127**:581-592.

* * * * * * * * * * * * * * * * * * * *

Questions:

1. True or False? Double minute chromosomes lack functional centromere.

2. Why do double minute chromosomes increase the potential for malignancy?

3. How are double minute chromosomes transmitted to daughter cells despite lacking the necessary structure to carry out this function?

Answers are at the end of the book.

2. - have diversified amplification of oncogenes.
   b/c - uneven segregation and accumulation

3. - have a lot of various behaviors and are closely assoc. w/ normal chromosomes.

- 9 -

The paper published in Nature *in April 1953 by Watson and Crick suggested a structure of DNA and hinted that the proposed base pairing sequence may be involved in a copying mechanism for genetic material. In May of that same year, Watson and Crick did indeed elucidate the self-replicating properties of DNA. Extensive discussion regarding the concept of self-duplication included the idea that a template could directly or indirectly duplicate genetic material. The details of DNA self-duplication were never explained at the molecular level, in terms of atoms and molecules, until now. This second paper included a brief discussion on the phenomenon of supercoiling in DNA and was the basis for future research in this area. The conclusions from this paper, much like the initial article published one month earlier, advanced the field of genetics by leaps and bounds.*

Genetical Implications of the Structure of Deoxyribonucleic Acid

by J.D. Watson and F.H.C. Crick

Nature, May 30, 1953

The importance of deoxyribonucleic acid (DNA) within living cells is undisputed. It is found in all dividing cells, largely if not entirely in the nucleus, where it is an essential constituent of the chromosomes. Many lines of evidence indicate that it is the carrier of a part of (if not all) the genetic specificity of the chromosomes and thus of the gene itself. Until now, however, no evidence has been presented to show how it might carry out the essential operation required of a genetic material, that of exact self-duplication.

We have recently proposed a structure[1] for the salt of deoxyribonucleic acid which, if correct, immediately suggests a mechanism for its self-duplication. X-ray evidence obtained by the workers at King's College, London[2], and presented at the same time, gives qualitative support to our structure and is incompatible with all previously proposed structures[3]. Though the structure will not be completely proved until a more extensive comparison has been made with the X-ray data, we now feel sufficient confidence in its general correctness to discuss its genetical implications. In doing so we are assuming that fibres of the salt of deoxyribonucleic acid are not artefacts arising in the method of preparation, since it has been shown by Wilkins and his co-workers that similar X-ray patterns are obtained from both the isolated fibres and certain intact biological materials such as sperm head and bacteriophage particles.

The chemical formula of deoxyribonucleic acid is now well established. The molecule is a very long chain, the backbone of which consists of a regular alternation of sugar and phosphate groups. To each sugar is attached a nitrogenous base, which can be of four different types. (We have considered 5-methyl cytosine to be equivalent to cytosine, since either can fit equally well into our structure.) Two of the possible bases — adenine and guanine — are purines, and the other two — thymine and cytosine — are pyrimidines. So far as is known, the sequence of bases along the chain is irregular. The monomer unit, consisting of phosphate, sugar and base, is known as a nucleotide.

The first feature of our structure which is of biological interest is that it consists not of one chain, but of two. These two chains are both coiled around a common fibre axis. It has often been assumed that since there was only one chain in the chemical formula there would only be one in the structural unit. However, the density, taken with the X-ray evidence[2], suggests very strongly that there are two.

The other biologically important feature is the manner in which the two chains are held together. This is done by hydrogen bonds between the bases. The bases are joined together in pairs, a single base from one chain being hydrogen-bonded to a single base from the other. The important point is that only certain pairs of bases will fit into the structure. One member of a pair must be a purine and the other a pyrimidine in order to bridge between the two chains. If a pair consisted of two purines, for example, there would not be room for it.

We believe that the bases will be present almost entirely in their most probable tautomeric forms. If this is true, the conditions for forming hydrogen bonds are more restrictive, and the only pairs of bases possible are:

adenine with thymine;
guanine with cytosine.

The way in which these are joined together is, for example, adenine can join on either chain; but when it does, its partner on the other chain must always be thymine.

This pairing is strongly supported by the recent analytical results[4], which show that for all sources of deoxyribonucleic acid examined the amount of adenine is close to the amount of thymine, and the amount of guanine close to the amount of cytosine, although the cross-ratio (the ratio of adenine to guanine) can vary from one source to another. Indeed, if the sequence of bases on one chain is irregular, it is difficult to explain these analytical results except by the sort of

pairing we have suggested.

The phosphate-sugar backbone of our model is completely regular, but any sequence of the pairs of bases can fit into the structure. It follows that in a long molecule many different permutations are possible, and it therefore seems likely that the precise sequence of the bases is the code which carries the genetical information. If the actual order of the bases on one of the pair of chains were given, one could write down the exact order of the bases on the other one, because of the specific pairing. Thus one chain is, as it were, the complement of the other, and it is this feature which suggests how the deoxyribonucleic acid molecule might duplicate itself.

Previous discussions of self-duplication have usually involved the concept of a template, or mould. Either the template was supposed to copy itself directly or it was to produce a 'negative', which in its turn was to act as a template and produce the original 'positive' once again. In no case has it been explained in detail how it would do this in terms of atoms and molecules.

Now our model for deoxyribonucleic acid is, in effect, a *pair* of templates, each of which is complementary to the other. We imagine that prior to duplication the hydrogen bonds are broken, and the two chains unwind and separate. Each chain then acts as a template for the formation on to itself of a new companion chain, so that eventually we shall have *two* pairs of chains, where we only had one before. Moreover, the sequence of the pairs of bases will have been duplicated exactly.

A study of our model suggests that this duplication could be done most simply if the single chain (or the relevant portion of it) takes up the helical configuration. We imagine that at this stage in the life of the cell, free nucleotides, strictly polynucleotide precursors, are available in quantity. From time to time the base of a free nucleotide will join up by hydrogen bonds to one of the bases on the chain already formed. We now postulate that the polymerization of these monomers to form a new chain is only possible if the resulting chain can form the proposed structure. This is plausible, because steric reasons would not allow nucleotides 'crystallized' on to the first chain to approach one another in such a way that they could be joined together into a new chain, unless they were those nucleotides which were necessary to form our structure. Whether a special enzyme is required to carry out the polymerization, or whether the single helical chain already formed acts effectively as an enzyme, remains to be seen.

Since the two chains in our model are intertwined, it is essential for them to untwist if they are to separate. As they make one complete turn around each other in 34 A., there will be about 150 turns per million molecular weight, so that whatever the precise structure of the chromosome a considerable amount of uncoiling would be necessary. It is well known from microscopic observation that much coiling and uncoiling occurs during mitosis, and though this is on a much larger scale it probably reflects similar processes on a molecular level. Although it is difficult at the moment to see how these processes occur without everything getting tangled, we do not feel that this objection will be insuperable.

Our structure, as described[1], is an open one. There is room between the pair of polynucleotide chains for a polypeptide chain to wind around the same helical axis. It may be significant that the distance between adjacent phosphorus atoms, 7·1 A., is close to the repeat of a fully extended polypeptide chain. We think it probable that in the sperm head, and in artificial nucleoproteins, the polypeptide chain occupies this position. The relative weakness of the second layer-line in the published X-ray pictures[3a, 4] is crudely compatible with such an idea. The function of the protein might well be to control the coiling and uncoiling, to assist in holding a single polynucleotide chain in a helical configuration, or some other non-specific function.

Our model suggests possible explanations for a number of other phenomena. For example, spontaneous mutation may be due to a base occasionally occurring in one of its less likely tautomeric forms. Again, the pairing between homologous chromosomes at meiosis may depend on pairing between specific bases. We shall discuss these ideas in detail elsewhere.

For the moment, the general scheme we have proposed for the reproduction of deoxyribonucleic acid must be regarded as speculative. Even if it is correct, it is clear from what we have said that much remains to be discovered before the picture of genetic duplication can be described in detail. What are the polynucleotide precursors? What makes the pair of chains unwind and separate? What is the precise role of the protein? Is the chromosome one long pair of deoxyribonucleic acid chains, or does it consist of patches of the acid joined together by protein?

Despite these uncertainties we feel that our proposed structure for deoxyribonucleic acid may help to solve one of the fundamental biological problems — the molecular basis of the template needed for genetic replication. The hypothesis we are suggesting is that the template is the pattern of bases formed by one chain of the deoxyribonucleic acid and that the gene contains a complementary pair of such templates.

One of us (J.D.W.) has been aided by a fellowship from the National Foundation for Infantile Paralysis (U.S.A.). ❑

References
[1] Watson, J.D., and Crick, F.H.C., *Nature*, **171**, 737 (1953).
[2] Wilkins, M.H.F., Stokes, A.R., and Wilson, H.R., *Nature*, **171**, 738 (1953). Franklin, R.E. and Gosling, R.G., *Nature*, **171**, 740 (1953).
[3] *(a)* Asbury, W.T., Symp. No. 1 Soc. Exp. Biol., **66** (1947). *(b)* Furberg, S., *Acta Chem. Scand.*, **6**, 634 (1952). *(c)* Pauling, L., and Corey, R.B., *Nature*, **171**, 346 (1953); *Proc. U.S. Nat. Acad. Sci.*, **39**, 84 (1953). *(d)* Fraser, R.D.B. (in preparation).
[4] Wilkins, M.H.F., and Randall, J.T., *Biochim. et Biophys. Acta*, **10**, 192 (1953).
[5] Chargaff, E., for references see Zamenhof, S., Brawerman, G., and Chargaff, E., *Biochim. et Biophys. Acta*, **9**, 402 (1952). Wyatt, G.R., *J. Gen Physiol.*, **36**, 201 (1952).

* * * * * * * * * * * * * * * * * * * *

Questions:

1. Which base pairs join together?

2. Why did the authors conclude that the salt of DNA was not an artifact arising from the method preparation but was actually important to their discovery?

3. How do the authors suggest that spontaneous mutations arise?

Answers are at the end of the book.

- 10 -

Prions and the havoc they wreak have been the bane of many cows' short existence in Great Britain. Prions are known to be responsible for the epidemic of "mad cow disease" or bovine spongiform encephalopathy, scrapie in sheep, and in humans, Creutzfeldt-Jakob disease. It was believed that these pesky proteins only existed to cause organisms neurological devastation. However, geneticists across the nation are discovering that these biological anomalies may play a role in healthy creatures as well as diseased ones. Studies on the yeast Saccharomyces cerevisiae, *demonstrate that rather than causing disease, prions are passing on genetic traits from one yeast generation to the next — a job considered the exclusive territory of DNA or its RNA counterpart. This article discusses the previous held belief that prions are infectious elements that carry their protein-distorting powers from one animal to the next and the new hypothesis that these "demons" may act as regulator proteins in tissues.*

Far From the Maddening Cows...

by Jonathan Knight

New Scientist, January 24, 1998

Don't throw away that gunk at the bottom of the tube — it could turn prions into the good guys after all.

Deep in the brains of some unfortunate British cows, changes are afoot. As normal proteins transform into sticky clumps, healthy calves become mad cows. Then dead ones. And although the strange "prion" proteins that are responsible have been cast as biological deviants because their only known function was to wreak neurological havoc, under the weight of new evidence that thinking is being transformed, too. Prions, it turns out, have a role to play, perhaps a big one, in healthy creatures as well as diseased ones.

The sea change in attitudes towards prions began when Reed Wickner, a geneticist with the National Institutes of Health near Washington DC, discovered prions in yeast. Rather than causing disease, Wickner's prions were passing on genetic traits from one yeast generation to the next, a job which for 50 years has been considered the exclusive providence of DNA.

The story might have ended right there — an astounding anomaly, to be relegated to a textbook footnote — except that there have been more sightings of these weird prion genes. Besides giving geneticists pause for thought, the new revelations are forcing biochemists to rethink their ideas about how proteins behave, and may compel developmental biologists to entertain new ideas about how embryo growth is regulated. And although prion genes have so far only put in an appearance in fungi, some say they will soon be popping up all over, even in humans.

"Every organism will have them," predicts Susan Liebman, a geneticist who studies yeast at the University of Illinois. "They will not be rare."

Curly-haired yeast

A heritable trait in the yeast *Saccharomyces cerevisiae*, whether passed on by DNA or anything else, is not as easy to spot as curly hair, big feet or blue eyes. Under the microscope, a yeast is a yeast is a blob. Yeasts do, however, metabolise chemicals in different ways. Most can't get their nitrogen from a chemical called ureidosuccinate whenever ammonia is available. But some yeasts, called [URE3] and pronounced "yuri-3," happily digest ureidosuccinate in the presence of ammonia. The [URE3] types of yeast show up in any population at the very low rate of about one in 100,000, and the way they pass on their ureidosuccinate-gobbling trait totally disobeys the most fundamental rules of inheritance.

✳Most of the time, yeasts reproduce by budding. But when the moment is right, they have sex. When yeast cells join, the union is total. The cells fuse, mixing everything: cytoplasm, proteins, nutrients, genes and cell walls. The fused parental cell then divides into four new daughter cells, distributing most of it component parts evenly — except for the genes. Each parent contains one

✳ each of the 16 yeast chromosomes, and each offspring gets one copy, chosen from the parental mix like a ticket from a raffle tumbler. That way if one parent has a trait that the other lacks, only half the offspring get it. It's a simple rule, but one that [URE3] seems to ignore.

Rather than appearing in just half of the progeny of a yeast coupling in which only one of the parents has the trait, [URE3] appears in all four daughters. Yeasts have a handful of other traits that disobey Mendel's laws, but these are carried on the genes of parasitic viruses or bacterial plasmids that infect all progeny when a yeast cell divides. No viruses or plasmids have ever been found that could account for [URE3]'s strange pattern of inheritance, or that of a second trait called [PSI], in which the cells make proteins that are too long. By 1989, when Wickner began working on [URE3], most researchers had given up looking for explanations to such riddles. "I felt like the den mother of these non-Mendelian elements, just because nobody else cared," says Wickner.

At that time on the other side of the Atlantic, the epidemic of "mad cow disease" or bovine spongiform encephalopathy was raging as furiously as the debate about its cause. At the core of the debate was a proposal from Stanley Prusiner

of the University of California in San Francisco Suggesting that proteins called prions caused a number of related brain diseases, including scrapie in sheep, mad cow disease, and, in humans, Creutzfeldt-Jakob disease. Prion proteins can flip from their regular shape to the disease-causing shape, and — in a domino effect — force proteins of the same composition to switch as well. The distorted proteins form insoluble clumps that gum up and eventually kill neurons, opening cavities in the brains of afflicted creatures.

The most controversial part of Prusiner's theory is that the prions are infectious, passing from one animal to the next, carrying their protein-distorting powers with them. Prusiner won one of last year's Nobel prizes for his work, but there are still those who argue that the "protein only" theory of infection is plain wrong. They say that infections can only be carried by DNA (or its RNA counterpart) contained in microorganisms like bacteria and viruses.

Wickner was inclined to believe Prusiner. He, too, was beginning to doubt DNA's omnipotence and, if prions worked as Prusiner said they did, they could explain how all the offspring of a yeast coupling can inherit a trait that only one parent had. But it was just a hunch, so he didn't involve any of his postdocs. "Why risk a young career? It was a long shot," says Wickner.

His idea went like this. A single distorted version of a prion protein could rapidly convert all the other proteins of the same composition in a yeast cell. When that cell fused with another, these prions in turn would sweep across the fused parental cell, distorting all the normal copies of the protein, and passing them on to the four daughter cells and every other descendant of the original cell down the generations. Wickner's candidate for such a protein was one already known to researchers called Ure2.

Yeast cells lacking the gene for the Ure2 protein behave just like [URE3] yeast, except that they pass on the trait in proper Mendelian fashion to just two progeny, suggesting that a malfunctioning Ure2 protein could be at the heart of the ureidosuccinate-gobbling trait in [URE3].

To test this hypothesis, Wickner crippled the

Ure2 gene in [URE3] yeast and let the cells grow for a few generations. If he was right, any inherited molecules of the distorted, presumably malfunctioning, form of Ure2 would break down as all proteins do in time. With no new Ure2 being made, the prions and the [URE3] trait would vanish.

To test whether he had successfully eliminated the distorted protein from the yeast colony, he reinserted the *Ure2* gene. As he had predicted, the Ure2 protein started working normally again, and the whole population of yeast was cured of [URE3]. Just as a cow must be infected with a prion "seed" to go mad, yeast needs a prion seed to create the [URE3] trait.

Instant sensation

The more tests Wickner did, the more it looked as if his weird genetic trait was carried by prions. Add guanidinium chloride to brain extracts from sheep with scrapie and the prion clumps break up, with the extract rendered noninfectious. Similarly, when Wickner added guanidinium chloride to his brews of [URE3] yeast, they were cured of the trait. What's more, says Wickner, [URE3] spontaneously reappeared in a small number of cured yeasts. A virus would have been gone for good.

When Wickner announced in *Science* in 1994 (vol 264, p 566) that yeast can use prions instead of genes to pass on heritable traits, it caused an instant sensation. "Nothing I've ever done in my life has brought forth such enthusiasm," says Wickner. One commentator proclaimed that the finding gave mammalian prions "additional respectability." Even researchers like pathologist Laura Manuelidis of Yale University, who are critical of the notion that prions are infectious in mammals, said that they found the story convincing.

Fellow yeast geneticist Susan Lindquist at the University of Chicago set out to confirm Wickner's findings. And a little over a year ago, she reported that, just like [URE3] yeast, [PSI] yeast cells are cured of their trait by temporarily stopping a gene, in this case for a protein called Sup35. She also tagged the Sup35 protein so it glowed green and

showed that under the microscope the luminescent protein formed clumps in [PSI] yeast. As expected, when those yeasts mated, they passed on the glowing green clumps to all four offspring.

But it was Liebman and Michael Ter-Avanesyan at the Cardiology Research Centre in Moscow who found evidence that prions are part of being a normal, healthy yeast, rather than a sign of some yeast equivalent of mad cow disease.

The Sup35 protein comes in two parts. Most of the Sup35 protein is needed for its regular job of halting the cell's synthesis of new proteins when they reach the proper length, but a small portion at the head of the Sup35 protein is entirely dispensable for that function. Yeasts lacking this extra bit are perfectly healthy, but, Ter-Avanesyan found, they can never be converted to [PSI]. He concluded that the dispensable portion was necessary for the non-Mendelian trait.

Liebman's lab confirmed the findings of Ter-Avanesyan and went one step further. For reasons that are still poorly understood, a blast of full-length normal Sup35 protein converts normal yeast to [PSI] yeast. Liebman found that a blast of the Sup35 small domain alone is sufficient to convert yeast into [PSI].

It turns out that the Ure2 protein also has a portion whose sole job is to enable the protein to switch between its two forms. Clearly, yeasts aren't producing prions by accident: dedicated parts of their proteins and genes make it possible for the proteins to flip from the soluble form to the clumping form.

Flip-flop

Last May, Lindquist published the first electron microscope photos ever of yeast prions, in this case Sup35, in their insoluble aggregates. The long, fibrous clumps looked almost identical to those seen in the brains of cows with mad cow disease and humans with Creutzfeldt-Jakob disease, adding even more weight to the idea that, far from being confined to the brains of sick mammals, prions happen in healthy yeast, and very likely other species, too.

"If you have something in yeast, and you have it in humans, you probably have it in everything," says Liebman. Peter Lansbury, a biochemist at the Harvard Medical School in Boston who studies mammalian prion diseases, agrees. Once people start searching for prions in other species, they'll find them, he predicts — as long as they know what to look for, that is.

One problem critics have always had with the idea of prions, whether they be passing on heritable traits or causing fatal brain diseases, is that proteins are supposed to be stiff, not flexible enough to flip from one state to another. "Because the first proteins described were very rigid, people have this idea that proteins are like rocks," Lansbury says. "But it's very clear now that proteins are flexible." So if you are hot on the trail of a prion in some species other than a yeast or a mad cow, don't ignore the fact that a prion protein will exist in more than one form.

To make the hunt slightly more challenging, prions, in their distorted form, tend to clump together. These clumps don't dissolve in water, so the prions get left behind, stuck to the bottom of the test tube, while the soluble form of the protein gets studied. "People spin down their samples," says Lansbury. "It's common practice. If it aggregates it's hard to study, so you throw it away."

Of course, the effort required to track down prions raises the issue of whether it's all worth it. What interesting things could the prions possibly be doing if they do turn out to exist in a wide range of living organisms? Prions are not independent entities like viruses or bacteria engaging in their own struggles to survive, but parts of living creatures. If their only role was to cause disease, evolution would have got rid of them. Lindquist suspects that [PSI] and [URE3] prions help whole yeast populations respond rapidly to changes in their environment. The [PSI] trait, for example, is more likely to occur spontaneously when a population is stressed, say by high temperature, suggesting that the trait may have something to do with helping yeast survive hard times. To check out that idea, Lindquist is now searching for genes that only function in [PSI] yeast.

Fred Cohen, who works with Prusiner at the University of California in San Francisco, also thinks that evolution will have found a way to make good use of prions, not least because of early hints that prions are far more common than anyone suspected. In a computer analysis of more than 100,000 theoretical proteins, Cohen and Prusiner have found that up to 3 per cent have two stable conformations, one of which can clump in true prion fashion. "I can't think of any detrimental phenomenon in biology that can't somehow be used to advantage in another setting," says Cohen.

The first case of prions being positively demonstrated to help an organism survive came in September, from French researchers working with a fungus called *Podospora anserina (Proceedings of the National Academy of Sciences*, vol 94, p 9773). The fungus grows in spreading mats on forest detritus. When one fungal mat encounters another, the cells at the edges fuse, much the way mating yeast cells do, mingling their contents. The trouble is that if one colony has a virus, both soon get infected. To minimise the damage, sometimes the fused cells sacrifice themselves for the good of the colony, dying and forming a barrier that stops the virus from spreading.

What determines whether the barrier forms is the prion-like transformations of a protein called Het-s, says Joel Begueret, a biologist at the Institute of Biochemistry and Cellular Genetics in Bordeaux. When a colony containing one form of the Het-s protein fuses with a colony containing the other, all the fungi quickly transform to a single form, as if a prion infection has swept through the colonies. The fused colony can then form the dead-cell blockage when it fuses with another colony.

Fine for fungi

Begueret has ruled out the possibility that a virus is responsible or that one colony is switching on genes in the other colony. What's more, if you add an artificial gene to the fungus to produce a blast of the normal form of the Het-s protein, the fungus converts to the form that makes the dead-cell blockage. "It sounds very prion-like," says Cohen.

Fine for fungi, but what about complex organisms such as humans? Do they use prions to pass on useful traits? No one is yet prepared to say that prions may be carrying heritable traits from human parents to their offspring, although theoretically it is possible. But prions may be involved in the elusive mechanism called imprinting that allows genes to behave differently depending on whether they are inherited from your mother or your father, says Lindquist (see "Where did you get your brains?" *New Scientist*, 3 May 1997, p 34).

Developmental biologists like Igor Dawid at the National Institutes of Health near Washington DC are happy to toy with the notion that prions may help the fertilised egg, an amorphous single-celled blob, become a multicellular organism with wings, livers, arms or antennas.

At some point in this transformation, the embryo cells decide whether to become liver, muscle or any other tissue. Once the decision is made, all their progeny have to stick with it. Yet all cells carry the same genetic code, so it's unclear how cells are passing on that information.

Prions could be responsible. Cells in different tissues manufacture different types of proteins, some of which keep the correct genes for that tissue turned on. If those regulatory proteins had the ability to spread their influence as prions do,

then dividing cells within a tissue would automatically know which genes to turn on. "No one would say that all such phenomena have to do with prions," says Dawid. "But some of them could, and that would be very exciting."

Certainly, when a group of cells in a developing animal or plant embryo sets off to become a particular type of tissue, large amounts of the regulatory proteins appear in a burst that later subsides. As fungi show, a burst of normal protein is often all it takes to trigger the appearance of its prion forms. Lindquist has even identified regulatory proteins that contain sequences resembling the bits of the yeast prions that allow them to switch between their two shapes, although she says its still too early to give specifics.

Meanwhile, if Lindquist, Lansbury and Liebman are right about the wide-ranging ramifications of the new "protein-only" genetics, throwing out uncooperative protein clumps may no longer be a wise option. Everyone from developmental biologists to yeast geneticists should probably start paying attention to the gunk stuck at the bottom of the tube. ❏

* * * * * * * * * * * * * * * * * * * *

Questions:

1. How do yeast normally reproduce and how many chromosomes do they contain? *budding* -16

2. What is the biggest argument against Prusiner's theory that prions are infectious? *- infections can only be carried by N.A. whether DNA or RNA contained in info*

3. True or False? Proteins are structurally stiff.

4. What is imprinting? *allows genes to behave diff. depending on whether they are inherited from mom or dad.* *- form of inheritance*

Answers are at the end of the book.

- prions in all creatures
- adv. to help yeast peopl's to adapt.
- may allow forms to respond to enrich. stress.
- prions allow pinprinting

- 11 -

The Human Genome Project is a worldwide research effort with the goal of analyzing the structure of human DNA and determining the location of the estimated 100,000 human genes. In parallel with this effort, the DNA of other organisms are being studied to provide the comparative information necessary for understanding the functioning of the human genome. The information generated by the human Genome Project is expected to be the source book for biomedical science in the next century and will be of enormous benefit to the field of medicine. It will help us to understand and eventually treat many of the more than 4,000 genetic diseases that afflict mankind, as well as the many multifactorial diseases in which genetic predisposition plays an important role. Through the use of computers to analyze biological data and by using robots, the Human Genome Project may be completed prior to its projected date of 2005.

A Long Molecular March

by David L. Wheeler

The Chronicle of Higher Education, **May 31, 1996**

Researchers start the painstaking process of sequencing the human genome.

Ten years ago, the idea of obtaining all of the molecular details of human DNA was little more than a gleam in the eyes of a few biologists.

Now, a robotic arm working seven days a week shuffles samples of human DNA through a series of analytical steps here at the Whitehead Institute for Biomedical Research. The work is part of the official beginning of a long molecular march through the human chromosomes.

One June 3, the robot's work — displayed in the four-letter code that signifies the sequence of the four chemicals on a stretch of human DNA — will be made available on the Internet (http://www-genome.wi.mit.edu) as part of a series of public releases of human-DNA data. The research is the result of a round of grants made last month by the National Center for Human Genome Research in Bethesda, Md., for a pilot phase of sequencing the human genome — all of the DNA on the 24 different chromosomes.

Scientists believe that having the complete DNA sequence will make it easier to search for the molecular errors that can cause disease, to understand the genetic basis of a healthy human body, and to compare human DNA with that of other organisms.

Grants to Six Institutions

The genome-research center, part of the National Institutes of Health, expects to spend $60-million over three years on the initial phase of sequencing. The money will be spread among six institutions. (Besides the Whitehead Institute, which is affiliated with the Massachusetts Institute of Technology, they are the Baylor College of Medicine, the Institute for Genomic Research, Stanford University, the University of Washington, and Washington.)

Biologists have already gathered a few of the details of human DNA as they search for the genetic flaws that cause some diseases. They have also arrived at the entire sequence for baker's yeast (Saccharomyces cerevisiae) and a few other microorganisms and are working on sequencing the genomes of a worm, a fruit fly, and a laboratory mouse.

The genetic data streaming across computer screens from such organisms is confirming what many biologists have long suspected: Nature cheats when creating new organisms. Instead of creating new genes, evolution adapts old ones to new purposes. A gene essential for creating the wing of the fruit fly, for example, is also essential for the development of human breast tissue.

"We've been talking evolution for a long time, but maybe we never believed it as much as we should have," says Eric S. Lander, head of the Center for Genome Research at the Whitehead Institute and a professor of biology at M.I.T.

No new genes have been created in the past half-billion years, he says. The biochemical mechanisms that control the activities of human cells, which match the control mechanisms in yeast cells, appeared on the planet along with yeast about 1.5 billion years ago.

At the genetic level, says Mr. Lander, life speaks a unified language. "In the past, people working in different organisms and different organs had nothing to say to each other. Now, as it turns out, it's all the same problem."

Some human cancers, for example, are caused by the inability of DNA to repair itself after it is disturbed. Scientists seeking ways to fix this flaw are studying yeast to understand how the repair process can go wrong.

The human-genome project, officially started in 1988, seeks to support medical research by gathering genetic information about humans and other organisms. In human DNA, the project has identified a series of molecular landmarks that will guide scientists in their search for genes on the chromosomes. The project is also setting up a "physical map," so researchers can reliably obtain a piece of DNA they want from any portion of the chromosomes. But the project's most ambitious goal has been sequencing the entire human genome.

A year ago at the Whitehead Institute, researchers looked at the number of technicians spread out over the laboratory performing the many steps used in sequencing. The researchers decided they could arrange all of the steps necessary in sequencing on a table and use a robotic arm sliding on a metal track to move the DNA around. Each piece of DNA is stuck to a tiny metal bead, which can be controlled by magnets. "We use fewer technicians but get 10 times as much work done," says Trevor L. Hawkins, head of the sequencing work at the institute.

DNA is composed of four chemicals — adenine, cytosine, guanine, and thymine — that are strung like beads on a two-strand necklace. Each chemical unit, or base, is interlocked with a partner in the other strand. The total sequence is expected to be about 3 billion base pairs long.

If the entire sequence of human DNA were printed out into books and the books were piled up, the stack would be 12 stories high, according to the National Center for Human Genome Research.

A New Era

Although the current stage of sequencing is billed as a preliminary phase, Mr. Lander is so certain that the genome project is headed quickly toward its conclusion that he is already talking about the "post-genome project era."

"There are those who felt that when the genome project was finally over, we could get back to biology as usual," he says. "But it's not going to happen."

The automation of laboratory work and the use of computers to analyze biological data have already sharply decreased the amount of time that biologists spend hovering over laboratory countertops, pipettes in hand. In the future, Mr. Lander says automation will make looking for disease genes more systematic and less dependent on luck. Gene-hunting now, he says, is like "going off into the bush with your machete and trying to find the gene and come back with it in one piece."

"Sometimes you succeed and come back with the gene and you're a big hero, and sometimes you get lost in the bush."

Benefits of Automation

The painstaking process of figuring out the structure and function of each gene will become increasingly automated, using a technological approach made possible by the genome project, Mr. Lander says. He envisions being able to toss cells into machines that will spit back information about which genes the cells use, and when. Biologists now estimate that about 100,000 genes exist, but that only a portion of those — maybe 10,000 — are used in each cell. Different cell types need different genes to do their jobs.

But scientists won't have to wait fort he genome project to be completed early in the next century for the post-genome era to begin, Mr. Lander says. Instead, highly technological approaches that analyze large amounts of cells or tissue and produce a large quantity of data will be applied almost immediately to such processes as understanding tumor cells.

He bases that prediction on his work with a National Cancer Institute committee that is considering how best to exploit the genomic project in the flight against cancer. Biologists are beginning to use automated methods to compare large numbers of tumor cells and normal cells to locate DNA mutations that cause disease, he says. Combining that information with patients' medical records could also show how each type of tumor cell responds to various forms of cancer treatment.

Mr. Lander says the technological future of the life sciences will increase the need for engineers, mathematicians, and computer scientists who also understand biology. Statistics and software are already important in trying to tease out the meaning of the long stretches of sequence in data bases. Only about 3 to 5 percent of the genome is actually used by cells to make proteins, the chief building blocks of the body. The rest of the genome regulates when genes are turned on and off— and some of the genome may have no function at all.

Despite the uncertainty about which portions are important, the scientists doing the sequencing need to get each part right. A flaw in just one of the base pairs in DNA can cause a disease, such as sickle-cell anemia. The length of sections that seem to be a meaningless repetition of matching sequences has turned out to affect the severity of some disorders.

Mr. Hawkins focuses his efforts on producing sequence data and making them public, not on analyzing the data or searching for genes. He has seen interesting sections of sequence go by. "I would love to keep it and take a look at it," he says, "but I have prevented myself from doing that."

The genome centers, he says, have a mandate to publish the data quickly. "If I sequence something on chromosome 17, I am not the world's expert on chromosome 17. Someone else out there has made a career out of this and is out there waiting for this information." ❑

* * * * * * * * * * * * * * * * * * * *

Questions:

1. How large is the human genome?

2. How much of the human genome is actually used by cells to make proteins?

3. Name the three main goals of the human-genome project.

Answers are at the end of the book.

- 12 -

Genetic engineering has long held the promise to revolutionize medicine and agriculture. For example, genetically engineered organisms could serve as "genetic factories" for the production of normal genes that could be inserted into humans with a serious heritable disease, such as cystic fibrosis. In addition, the production of genetically superior crops that have the capability of surviving in less than perfect conditions may help third world countries where millions are dying each year from starvation. None of this was possible on a large scale until the cloning of Dolly. The investigators that created Dolly used a novel approach to cloning. This group transferred nuclei of lab-grown cells from very young sheep fetuses, into eggs that had their own genetic material removed, triggering the development of an embryo, and eventually a whole sheet — basically a clone of the animal that donated the original cells. Although this new way of streamlining genetic engineering has some clear benefits, many questions and controversies surround this new technique.

We Ask They Answer

by Phillip Cohen and Rachel Nowak

New Scientist, May 9, 1998

One sheep looks much like another, so why clone them?

As Ian Wilmut will tell anyone who cares to listen, the main aim of his team at the Roslin Institute in Scotland was to transform the genetic engineering of farm animals (sheep first, then cows) from a hit-and-miss experimental procedure into a robust technology. Cloning was just a welcome by-product; Dolly, the extra cheese on the pizza.

Genetic engineering has long promised to revolutionise medicine and agriculture. A few sheep and cows, for instance, have already been engineered to produce human proteins such as clotting factors in their milk. And in future, their cousins could have their genes altered so that they develop human genetic diseases like cystic fibrosis, or even heart disease.

A sheep with cystic fibrosis sounds gruesome, but Wilmut believes the advantages would outweigh the cost in animal suffering. Cystic fibrosis is one of the most common, serious heritable diseases. At the moment, drugs to treat the disease can only be tested in mice. Sheep lungs are far more like out own.

A more distant possibility is that genetic engineering could turn out farm animals with organs that look human, immunologically speaking, providing an endless stock of organs for transplant. But the idea is controversial because of the risk of passing diseases from animals to humans. On the agricultural front, cows could be engineered so that they produce lots of milk without being susceptible to mastitis, a painful inflammation of the udder. Then there's the possibility of tweaking genes to make meat lean, or to get low-fat milk straight from the cow.

None of this was feasible on a large scale until the Roslin team's breakthrough. Before then, the only way to genetically engineer sheep and cows was to inject a gene directly into a newly fertilised egg. This is horribly inefficient. You have to inject thousands of fertilised eggs to get one animal that makes it to adulthood with the new gene active in the correct tissues. And even if you succeed once, there is no guarantee that you will be able to do it again.

To get round those problems, Wilmut and his colleagues wanted to be able to do their genetic manipulations in sheep cells growing in a flask. This makes genetic engineering far more efficient because you can insert and remove genes, or even create tiny mutations in the DNA, and then allow the cell population to grow. From this population you can take out a few cells to check that you have changed the genes correctly, and still have thousands of identical copies left to play with. But it does mean you need a way of turning the cells back into sheep. This is where the Roslin team made their biggest contribution. They transferred the nuclei of lab-grown cells from very young sheep embryos, and then from older sheep fetuses, into eggs that had had their own genetic material removed, triggering the development of an embryo and eventually a whole sheep — that is, a clone of the animal that donated the original cells.

Dolly came along when they tried the same thing using adult sheep cells grown in a flask. And since Dolly has become Polly — proof positive that you can combine genetic engineering (she'll secrete a human protein in her milk) and the cloning of lab-grown fetal sheep cells.

Dolly also raises the possibility of a whole new way of streamlining genetic engineering. Once wrinkles in the technique have been ironed out, researchers will be able to clone directly from genetically modified adult farm animals that have already proved their value as biofactories or food producers. — Rachel Nowak

Why start with an egg to make a clone of an adult?

"The most remarkable substance on the planet" is how developmental biologist Keith Latham of Temple University School of Medicine in Philadelphia describes the egg cytoplasm, the goo surrounding the nucleus.

Latham's enthusiasm is understandable. After all, Dolly is alive and well because an egg reprogrammed a cell designed for the very adult task of being part of an udder.

The resurrecting proteins of the egg's cytoplasm — most of them still unidentified — are there to prepare the sperm's nucleus for its union with the egg's genetic material, and to orchestrate the first cell divisions that turn the embryo into a ball of eight or 16 cells. Only at that point do the new embryo's genes take the helm.

As the embryo grows further and cells start to specialise, genes that are needed only for the early stages of development become cloaked in proteins called histones and are turned off, while the genes a particular cell needs to perform its adult functions are turned on by proteins called transcription factors. To send an udder cell back to its unspecialised state, the egg had to turn off the adult genes and reactivate the embryonic ones.

"In reprogramming donor cells, the egg is mimicking what it does to the sperm nucleus," says Keith Campbell, one of Dolly's creators, who is based at PPL Therapeutics in Edinburgh. The sperm, too, is a specialised cell, although its task is a pretty simple one. It's designed for the express purpose of carrying the father's genes from testes to egg, and to do that the adult genes necessary for the trip must be turned on.

To reprogram sperm genes, the egg cytoplasm contains large amounts of a protein called MPF that dissolves nuclear membranes, condenses chromosomes, and loosens the grips of transcription factors. It contains another enzyme that clips specific bonds between proteins known as protamines, which help package the sperm chromosomes. Immersion in the huge bath of cytoplasm — the egg is a very big cell — also helps to slough the proteins from the chromosomes.

Remarkably, mouse cells come equipped with a means of removing histones too, according to as yet unpublished research by developmental biologist Lawrence Smith of the University of Montreal in Canada. For cloning researchers that might turn out to be an important quirk of egg design — sperm may not contain histones, but the cells used in cloning do. — Phillip Cohen

Could somebody clone you?

Suppose someone wants to have a little baby copy of you (technology permitting). Maybe it's an old boyfriend or ex-wife who is still smitten, or some scary new-style stalker. Whoever they are, they would have little problem finding some of your precious cells to try to clone from — and the chances are you'd never even know it.

"The average person doesn't realise they leave tissue around all the time," says Robin Alta Charo, a law professor at the University of Wisconsin, Madison. If you have ever given blood, had a biopsy or surgery, then a piece of you may still be tucked away clearly labeled in a pathologist's freezer. In many US states and European countries, blood is routinely taken from newborns for genetic screening and sometimes stored for decades.

But for one-stop shopping for a well-documented fragment of human tissue, it's hard to beat the Armed Forces Institute of Pathology in Washington DC, a tissue bank used by pathologists and forensic scientists. It holds about 92 million tissue bits from civilians and military personnel around the world, and millions more are added each year. Increasingly such samples are given or sold to academics and biotechnology companies for genetics research.

And if the would-be cloners can't find a scientific institution to front for them, they could try surreptitiously swabbing your wine glass or plucking some of your hairs — that might just provide enough cells to generate a clone. — Phillip Cohen

Isn't there a law against it?

If you discovered you were the victim of body banditry, calling the police wouldn't necessarily help, says Lori Andrews of the Chicago-Kent College of Law in Illinois. Cloning humans is banned in only a few places — Britain, but not in most states in the US and Australia. What's more, "there isn't much law that lets us control our tissues once they are outside our bodies," says Andrews. She cites, for example, the 1980 lawsuit brought by John Moore, who discovered that researchers had used a sample of his spleen to make a new drug worth billions of dollars. The California Supreme Court found that Moore had no legal rights to share in that bounty.

Even your right to assume parentage of your cloned offspring would be ambiguous at best. In most European countries, the law defines the mother of a child as the woman who has carried the pregnancy; the father is variously defined as the partner of the woman who gestates the child or the sperm provider. In the US, some states rely on genetic tests to establish parenthood in case of disputes.

But if a child was secretly cloned, say, from your blood, you will have contributed neither sperm nor eggs, and you won't have been involved in the gestation. Routine maternity or paternity genetic testing would only add to the confusion by showing that you are too closely related to be a parent, and were instead a twin.

Legislators are unlikely to be in any hurry to address the dilemmas of human cloning for fear of tacitly condoning a controversial idea. But eventually the law will need to deal with the new realities, says Alta Charo. "To protect people's rights, we'll have to revisit the logic of how we view our bodies and relationships like mother, father, sister, brother. Right now we are kind of adrift." — Phillip Cohen

Would it be weird to be a clone of your Mum or Dad?

For identical twins life is bittersweet. They have enviably close relationships. Yet they also have trouble "developing an identity, developing a sense of their own agency in the world," says Ricardo Ainslie, a psychologist at the University of Texas in Austin who studies twins. "It's an issue that most of us take for granted. But identical twins grapple with it all the time."

In an effort to forge their identities, twins can pigeonhole themselves — one may become overly artsy, the other too practical. "It can rob you of opportunities to use your talents," Ainslie says. Cloned children could have similar identity conflicts with their parent, he says, and one possible outcome is a double dose of teenage rebellion as the child attempts to establish his or her own identity.

Of course, a lot depends on the parent. "Parenting is partly narcissistic," says Ainslie. "It allows us to be selfless, but it also means that you sometimes insert your own interests in the place of the child." From pressuring their children to join the family firm, to disapproval if they support a rival football team, normal parents' narcissistic tendencies could be amplified by clonehood, Ainslie fears. The child could "feel like nothing but an extension of the parent's needs."

Let's not get carried away, says Renee Garfinkel, a psychologist at the Adoption Studies Institute in Washington DC. Every factor that helps a child create a healthy sense of self has an impact, from birth order and family style to a child's talents, physical appearance, and whether or not he or she is clone. "But no one factor is ever decisive," she says.

Garfinkel also argues that fears of sexual interest from the nongenetic parent are unfounded. Sure, the child will bear an uncanny physical resemblance to the other parent as he or she reaches adulthood, but this counts for little. According to studies of adopted children and unrelated children raised together, the aversion to incest is created during rearing and is not dependent on who you are genetically related to or on physical appearance, Garfinkel says. — Rachel Nowak

What about a Manhattan Project for human cloning?

Picture this. Dolly's birth is announced and the world cheers. There are no doubts, no wariness on the part of researchers, religious leaders, politicians, or the public. They speak with one voice: we want human clones!

In response, nations unite to create a lavishly funded International Institute on Cloning, charged with copying adult humans as quickly and safely as possible. Its first priority is to improve cloning efficiently. Dolly took more than 400 attempts,

including 28 miscarriages. Such a high failure rate is unacceptable for humans.

One theory is that adult cells are more set in their ways than, say, the cells of a fetus or an embryo, which are easier to clone. Since the cytoplasm of the egg does the reprogramming magic, IIC cell biologists try passing the nucleus of an adult cell through not one egg, but two. They also try fusing nuclei with less mature eggs, which typically stall at the reprogramming step, giving extra hours in which to remould the nucleus.

A second explanation for the inefficiency of current cloning techniques is that most adult cells are damaged beyond the egg's ability to repair them. As a cell ages it collects genetic scars: its chromosome ends shorten, and DNA mutations accumulate. The damage may not be obvious. An udder cell would function perfectly well no matter how mangled its heart genes have become, because they are switched off. But if the cell was used in cloning, the embryo would develop cardiovascular problems and be aborted.

To get to grips as quickly as possible with the problem of genetic damage, IIC geneticists focus on mice, reasoning that their genetic make-up is well known, and it's possible to study several generations in a matter of months. With those mice, they identify the types of cells that are least prone to age-related damage, and develop efficient new ways to screen for genetic damage, and even to reverse it.

In another wing of the sprawling IIC, reproductive technologists start on the human end of the project. They offer women made infertile by defects in their egg cytoplasm the option of having their egg's nuclear material transferred into a donor egg with a healthy cytoplasm and then fertilised.

Meanwhile, IIC tissue engineers transfer the nuclei of cells taken from patients with Parkinson's disease into eggs that have had their own nuclei removed, so creating an embryo clone. Every cell in the very young embryo is pluripotent — capable of generating any type of adult cell. They put the embryo cells in a flask and try adding different nutrients and growth factors at different times in an attempt to grow new nerve cells. Once they succeed, they will have a stock of healthy nerve cells that are a perfect match for the Parkinson's patients who donated the originals — ready to be transplanted back.

So far the IIC team hasn't tried to create a human clone that makes it to adulthood. But they are getting a feel for nuclear transfer with human cells, the technique that underpins cloning. It will stand them in good stead come the fateful day when they make their first attempt to copy a human. — Phillip Cohen

But real life's not like that...is it?

It's not as different as you might think. Despite widespread misgivings about human cloning, researchers are pursuing the very same avenues of investigation as the fictional IIC. That's because every technique that brings human cloning closer also holds great promise for medicine and agriculture. In the past year, calves have been born that were created from fetal cell nuclei that had been passed through two eggs or inserted into a younger stalled egg.

Already infertile women have been treated by injecting healthy cytoplasm into their eggs (*The Lancet*, vol 350, p 186). Now, there are plans afoot to use nuclear transfer to treat infertility caused by problems in the egg cytoplasm. Meanwhile, Parkinson's patients routinely receive transplants of nerve cells from fetuses. Using cloning to grow those or other types of human cells for transplant is a much talked about prospect, and the concept has just been tested using cloned cow cells in rats see this month's *Nature Medicine*, vol 4, p 569).

Don Wolf of the Oregon Regional Primate Research Center in Beaverton, who is working on primate cloning for AIDS research, says that human cloning might be feasible within as little as five years. "Step by step," he says, "the cloning industry will get better at what they do." — Phillip Cohen ❑

Birth Announcements

▸ 1960s — 1970s

Frogs are cloned from tadpole cells and from frog cells by John Gurdon at the University of Cambridge. The latter proves so difficult — none make it to froghood — that most researchers are convinced that cloning adult animals is doomed to fail. The trend of naming animal clones has not yet caught on.

▸ 1981

First mammals are cloned — or are they? Karl Illmensee and Peter Hoppe at the University of Geneva claim to have cloned mice from the cells of very young embryos. Later experiments cast severe doubt on the claim.

▸ 1986-7

Sheep and cows are cloned from unspecialised early embryo cells by Steen Willadsen at the AFRC Institute of Animal Physiology in Cambridge and Neal First at the University of Wisconsin, Madison. This time it's indisputable.

▸ 1995

Morag and Megan are born at the Roslin Institute. The two sheep are the first clones created by nuclear transfer from partially specialised embryo cells grown in a flask.

▸ 1997 February

Dolly's birth is announced. She's cloned from a fully specialised adult udder cell and becomes the most famous sheep of all time.

▸ 1997 March

Neti and Ditto, two rhesus monkeys are born at the Oregon Regional Primate Research Center in Beaverton. They are the only primate clones ever, says Don Wolf, their creator. They come from unspecialised embryo cells.

▸ 1997 Summer

For the first time, cloning technology is used with genetic engineering. Polly is cloned by the Dolly team from fetal cells that have been engineered to contain the genes for a human clotting factor.

▸ 1997 August

Six month-old Gene is announced by ABS Global Inc, a biotech company in DeForest Wisconsin. He's the first cow to be cloned from a specialised fetal cell.

▸ 1998 February

Marguerite, a clone of a muscle cell from a calf fetus, is born at the French National Agricultural Research Agency near Paris. She dies from an infection six weeks later.

▸ 1998, 13 April, 4 am

Dolly delivers a baby, Bonnie, proving that despite her unusual origins she can still reproduce.

* * * * * * * * * * * * * * * * * * * *

Questions:

1. What is the role of the proteins in the egg cytoplasm?

2. What are the theories why cloning efficiently is so poor?

3. What is pluripotent?

Answers are at the end of the book.

Part Three

Chromosomes: Structure, Function, and Replication

- 13 -

Centromeres play a crucial role in accurate segregation of daughter cells in mitosis and meiosis. Chromosomes that lack centromeric DNA (structurally acentric chromosomes) are usually not inherited in mitosis and meiosis. However, studies have demonstrated that structurally acentric mini-chromosomes display efficient mitotic and meiotic transmission despite their small size and lack of centromeric DNA. This review discusses how researchers have now shown that other DNA sequences bind to a centromere-specific protein that associates with the spindle poles in anaphase. Using the fruit fly as the experimental model organism, the researchers found that a tiny piece of the fruit fly chromosome breaks off and grows a new centromere that allows the fragment to be copied like a mini-chromosome. If scientists can confirm what causes the normal DNA to take on the centromere's task, this may lead to the development of man-made chromosomes that can be used for gene therapy.

The Great Divide

by Nell Boyce

New Scientist, January 1998

Even chopped up chromosomes can replicate like a dream.

In the intricately choreographed process of cell division, strips of DNA called centromeres, where dividing chromosomes split, play a vital part. Now researchers have shown that other DNA sequences can somehow learn the centromere's dance. This may help scientists create artificial chromosomes for use in gene therapy.

Gary Karpen and his colleagues at the Salk Institute in La Jolla, California, witnessed a tiny piece of a fruit fly chromosome breaking off and growing a new centromere that allows the fragment to be copied like a minichromosome. In the past, doctors have come across rare cases of minichromosomes in patients, but the origin of newly sprouted centromeres remained a mystery.

Normally, before a cell divides, each chromosome is copied and the pairs line up in the cell's centre. Protein fibres from each end of the cell grab the chromosomes at the centromere. The two chromosome copies then break apart and move along the fibres to opposite sides of the cell, which splits down the middle.

Many researchers believed that the key to the centromere's function was that it always had the same characteristic DNA sequence. But it turned out that the sequence varies widely — not only between species but also between a single organism's chromosomes.

To find out more about how the centromere works, Karpen's team irradiated a fruit fly X chromosome to break it apart and rearrange it so that the centromere was close to one end. Further irradiation broke off a tiny piece of the tip, only about 0.05 per cent of the whole chromosome, leaving the normal centromere behind.

Remarkably, a copy of the tip was reliably passed on to daughter cells. "That tip was surprisingly transmittable through cell division," Karpen says. Several tests suggested that it must have "sprouted" its own centromere, his team reports in this month's *Nature Genetics* (vol. 18, p. 30).

Karpen suspects that the DNA tip developed a centromere because it had been moved close to a centromere on the X chromosome. A factor from the centromere may possibly spread out over nearby DNA, which can then be retained, making the DNA function like a centromere when it is isolated.

Alternatively, other DNA might contain dormant centromeres that are "switched off" by the working one. When the DNA is snipped off and the inhibition stops, the dormant centromere goes into action. But Karpen thinks this idea is less likely. "If you cut the normal end of X off, it does not act like that," he points out.

"It's quite a big change in the way that people are thinking about the centromere." says Chris Tyler-Smith of Oxford University, who also studies centromeres. If scientists can confirm what induces normal DNA to take on the centromere's task, this may lead to artificial human chromosomes that can replicate fast enough to be practical in gene therapy. ❏

* * * * * * * * * * * * * * * * * * * *

Questions:

1. During cell division, where do chromosomes split?

2. True or False? The DNA sequences of centromeres are always the same within species.

3. After the researchers irradiated the fruit fly X chromosome, causing a piece to break off, how did a copy of that missing tip get passed on to the daughter cells?

Answers are at the end of the book.

- 14 -

Fluorescence in situ *hybridization (FISH) is one of the most versatile and useful technologies available to the genetic's community today. Almost any stretch of genomic DNA may be appropriate for FISH analysis once a large DNA clone is available which contains the locus. The technique of in situ hybridization was first described as long ago as 1969 (John et al, 1969, Gall and Pardue,1969). However, for several years its use was limited by the availability of nucleic acid probes, which had to be radioactively labeled (H3 or I125). A number of developments in the 1980s overcame this problem and have led to much more widespread use of the technique. Recombinant DNA techniques meant that specific probes could be produced in large quantities, and subsequent developments such as automated oligonucleotide synthesis, non-radioactive labeling and PCR have helped to make it a routine technique in many laboratories. However, the application of in situ hybridization is still marginal compared to more widely used techniques such as PCR and Southern Blotting, probably because of the relative lack of standard protocols, dedicated instrumentation, and reagent kits. Applications of in situ hybridization fall into three main areas: (1) identification and localization of viral nucleic acids; (2) studies of mRNA expression; and (3) chromosome analysis and gene mapping. FISH analysis has now been redefined and the brilliance of this technique is shining through in its true colors. Maybe Dr. Seuss was predicting the future of karyotype analysis when he said, "One FISH, two FISH, blue FISH, black FISH, blue FISH, old FISH, new FISH. Say, what a lot of FISH there are."*

One FISH, two FISH, red FISH, blue FISH

by Michelle M. Le Beau

Nature Genetics, April 12, 1996

Chromosomal abnormalities are a leading cause of genetic diseases, including congenital disorders and acquired diseases such as cancer. Cytogenetic analysis by conventional chromosomal banding techniques, although highly precise, requires skilled personnel and is labour intensive, time consuming and expensive. Automated karyotyping is useful for some diagnostic applications, such as prenatal diagnosis, but, because the available computer algorithms are not sufficiently robust, it is ineffective in analysing the complex karyotypes characteristic of many human tumours. These factors have led investigators to seek alternative methods for identifying chromosomal abnormalities. The technique of fluorescence *in situ* hybridization (FISH) is one such method.

Few scientists will ever forget the first time they peered into a microscope and visualized the brilliant colours of a FISH study. For cytogeneticists, the experience is particularly memorable; the excitement of confirming that we were correct in our interpretation of a particularly complex abnormality, or the humbling experience of realizing our own limitations in defining abnormal chromosomes can scarcely be forgotten. At present, FISH is used largely to detect *known* chromosomal abnormalities using specific DNA probes. Unlike karyotype analysis, FISH has not become a screening test for chromosomal abnormalities, because of the inability to distinguish spectrally overlapping fluorochromes and, hence, to discriminate each human chromosome in a different colour. The technological advance described by Speicher et al.[1] has overcome these limitations, and may redefine how we perform diagnostic and research cytogenetic analysis in the near future.

The high sensitivity and specificity of FISH, the speed with which the method can be accomplished (virtually overnight), and the ability to provide information on a single-cell level have made a FISH a powerful tool. One of the earliest limitations of FISH was the inability to distinguish more than one target sequence. The introduction of multi-colour FISH allowed the detection of more than one target sequence, but was still hampered by the limited number of fluorochromes available[2]. In 1990, Nederlof et al.[3] described a method for detecting more than three target DNA sequences using only three fluorochromes. Using this technique, referred to as 'combinatorial labelling'[2-4] or 'ratio-labelling'[5], a probe can be labelled with more than one hapten (or fluorochrome-conjugated nucleotide) in varying ratios and detected with more than one fluorochrome. This strategy was demonstrated in 1992 by David Ward and his colleagues at Yale University, who used combinatorial labelling with three fluorochromes along with digital imaging microscopy to detect 7 DNA probes simultaneously in metaphase cells[4]. The number of useful combinations of N fluorochromes is 2^N-1; thus, increasing the number of fluorochromes to four allows the detection of 15 different target sequences, 31 combinations for five fluorochromes, and so on. By 1993, the goal of visualizing the 22 human autosomes and 2 sex chromosomes in 24 different colours seemed close at hand, but the next few years proved that this was a formidable challenge indeed.

Now, in a beautiful (scientifically and literally) study reported in this issue, Speicher et al.[1] have achieved this elusive goal. These investigators have developed epifluorescence filter sets and computer software to detect and discriminate 27 different DNA probes hybridized simultaneously. For cytogenetic analysis, a pool of human chromosome painting probes[6], each labelled (or detected) with a different fluorochrome combination, was hybridized to metaphase chromosomes. The result is a dazzling colour display of the human chromosomal complement! The major technological advances are (i) the selection and testing of filter sets based on excitation and emission contrast ratios that achieve >90% discrimination of each fluorochrome from its two nearest spectral neighbours, and (ii) the development of computer software for the analysis of the spectral signature of each probe. Furthermore, the investigators have demonstrated that the technique (termed multiplex FISH) can be used to detect simple and complex chromosomal abnormalities in a variety of clinical specimens from individuals with constitutional abnormalities or from tumour samples. In some cases, multiplex FISH was helpful in resolving abnormalities that were not defined precisely by conventional cytogenetic analysis.

Schröck et al. (manuscript submitted) have also reached this milestone by a novel approach that combines Fourier spectroscopy, CCD-imaging and optical microscopy to distinguish multiple spectrally overlapping probes. In contrast to the approach of Speicher et al.[1], which uses a filter wheel to capture separate images for each fluorochrome, their approach uses a triple-pass filter set that allows all dyes to be excited and measured together without image shift. The computer software employs a spectral-based classification algorithm that identifies the various spectra in the image and assigns each a pseudo-colour.

Now that colour karyotyping is a reality, what can we do with it (Table 1)? First, multiplex FISH will be applicable to the identification of numerical abnormalities, such as simple gains and losses of whole chromosomes, as well as some, but not all, structural rearrangements. Simple and complex translocations and relatively large deletions will be identified with ease. One plausible scenario is that multiplex FISH will be used as a screening test to identify numerical abnormalities or rearranged chromosomes. The identification of the chromosomal breakpoints of translocations can then be accomplished using fractional length analysis and chromosome band conversion charts[7].

However, it is likely that the greatest precision will be obtained by combining conventional analysis of banded metaphase cells with fractional length determinations or the hybridization of specific single copy probes to define the precise breakpoints and composition of the rearrangement. Future advances, such as the spectral resolution of one more fluorochrome, will be required for the simultaneous use of hybridization banding with probes complementary to LINE or SINE (*Alu*) sequences, or the hybridization of chromosome arm-specific probes for the entire genome[8]; these advances will add a new level of precision.

With respect to cancer cytogenetics, the application of multiplex FISH may result in the recognition of *new* recurring abnormalities in diseases that have been well-characterized, such as leukaemias and lymphomas, or in tumours that have been less amenable to conventional cytogenetic studies. A recent example in the haematologic malignant diseases is illustrative, and it has come as quite a surprise to cytogeneticists that a t(12;21)(p13;q22) is the most common abnormality in childhood B cell acute lymphoblastic leukaemia, occurring in at least 25% of patients[9]. This abnormality was not detected by the analysis of banded chromosomes, because the banding pattern of the involved chromosome bands and the size of the translocated material was so similar. The question remains of how many other undetected recurring rearrangements there are. Multiplex FISH provides a powerful tool to answer this question as well as the question of how many leukaemia patients truly have a normal karyotype (most laboratories report a normal karyotype in 15–20% of cases).

Other exciting applications of multiplex FISH are comparative cytogenetic studies to examine phylogeny and analyses of chromosome domains and their interrelationship in the interphase nucleus. Schröck *et al.* (manuscript submitted) used multiplex FISH of human chromosome painting probes to reconstruct the karyotype of a gibbon species; thus, the method could be applied to examine chromosome evolution or the degree of homology of various sequences. This application may be particularly useful in the agriculture industry or in plant biology, where questions of speciation are an important focus of current research efforts.

There are some existing limitations of karyotyping by multiplex FISH. Multiplex FISH of painting probes will not be useful in the detection of paracentric inversions, some pericentric inversions, insertions involving a single chromosome arm, small duplications and small deletions as in the clinically relevant microdeletion syndromes. The sensitivity in detecting cryptic telomeric translocations is untested. Designating the breakpoints of some rearrangements, for example, interstitial deletions, will only be possible by conventional cytogenetic analysis. With respect to double minute chromosomes and homogeneously staining regions (HSR), the cytogenetic manifestations of gene amplification, multiplex FISH analysis may be helpful in identifying the chromosomal origin of the former abnormality. HSRs may be complex in composition, and multiplex FISH may identify the origin of the chromosomal material contributing to an amplified segment[10]. In some cases, however, simple HSRs may appear as an insertion of chromosomal material into a recipient chromosome, but multiplex FISH will not reveal the uniform staining pattern that would identify it as an HSR. Some of the limitations mentioned above can be addressed by using FISH probes that target a specific disease related region, such as probes within microdeletions. It is likely that multiplex FISH will be strongest when used in a complementary manner with conventional cytogentic analysis. Thus, multiplex FISH can identify the gross abnormalities, and more conventional FISH or cytogenetic analysis can define the abnormality more precisely.

The technological achievements described above are stunning, yet it is difficult to predict how easily this technology will be transferred from one research laboratory to another and, more importantly, from the research laboratory to a clinical diagnostic setting. As described by Ledbetter[11], the initial introduction of FISH methodology in the late 1980s was followed by a great deal of excitement over the new technology,

---
Table 1 Applications of multiplex FISH

Clinical applications
  Medical genetics and cancer diagnostics:
  - Detection of numerical and structural chromosomal abnormalities
  - Identification of marker chromosomes (rearranged chromosomes of uncertain origin)

  Cancer diagnostics:
  - Monitoring the effects of therapy, and detection of minimal residual disease
  - Detection of early relapse
  - Identification of the origin of bone marrow cells following bone marrow transplantation
  - Identification of the lineage of neoplastic cells
  - Examination of the karyotypic pattern of non-dividing or interphase cells
  - Detection of gene amplification

Research applications
  - Molecular analysis of chromosomal abnormalities in human tumours
  - Multiplex chromosomal localization of genes and DNA sequences
  - Preparation of physical maps (including determining the order of and distance between sequences)
  - Characterization of somatic cell hybrids
  - Facilitate the identification of chromosomes containing genes for desirable traits in hybrid plants or crops and isolation of the gene(s)
  - Phylogenetic studies to determine the location and extent of synteny between different species
  - Detection of amplified genes
  - Detection of viral sequences in cells
  - Qualitative and quantitative assessment of mRNA or proteins on a cell-by-cell basis
  - Analysis of RNA processing and transport
  - Examination of the organization of chromosomal domains or multi-gene families in interphase nuclei as a function of tissue type, developmental status, cell cycle stage or disease state (3D reconstruction)
  - Analysis of chromosome aberrations in genetic toxicology studies

---

then by the phase of 'testing' FISH in a research and diagnostic setting and, finally, a 'disappointment' phase with the recognition of the pitfalls and limitations. FISH has been applied widely to research applications[12], and has had a tremendous impact in many disciplines[13], such as cancer genetics[14] and human genome mapping[12] to name a few.

Nonetheless, FISH has not had the impact on clinical diagnostics that many cytogeneticists and geneticists predicted 5 years ago. The reasons for this are manifold. First, FISH can be costly. For multi-colour applications, an imaging system and image analysis programs are typically required. Together with the costs of a microscope equipped for fluorescence, the start-up costs can far exceed $150,000. Probes and kits are also significant expenses. Second, the criteria for acceptable performance (sensitivity and specificity) have not been established. Third, regulatory agencies have been slow to approve FISH probes for clinical use. The fourth, and perhaps most important reason, is that FISH does not circumvent the need for a complete cytogenetic analysis, since the only abnormalities that are detected are those for which specific probes are employed. For example, a probe set for chromosomes 13, 18 and 21, and for the X and Y chromosome will only detect between 80–95% of the unbalanced chromosome complements that may result in liveborns with phenotypic abnormalities [11, 15]. In this regard, the development of multiplex FISH makes screening the entire chromosomal complement with a single test a viable option. Although many factors, including the nature of the health care system, influence the decision as to what level of sensitivity is acceptable for a diagnostic test, it will be extremely difficult in the United States to adopt a prenatal diagnostic test that has less accuracy than those currently available. Other factors that may have had a negative impact on the introduction of FISH technology into widespread usage are low sample volume for some laboratories, the perception that FISH is a 'molecular' test and the unwillingness to adapt to fluorescence microscopy.

Will FISH ever replace conventional cytogenetic analysis in pre- and post-natal diagnosis, or cancer diagnostics? Until recently, the answer was a resounding 'No.' With the new capabilities provided by multiplex FISH, such as the ability to analyse all of the human chromosomes in a single hybridization experiment, this answer has substantially less conviction. An important concern, however, is that the multiplex FISH technology requires additional equipment and computer analysis programs, resulting in a widening gap between scientists at the leading edge of technology development and the practicing cytogeneticist or genetics research scientist.

The challenge for the future is to bring these interests together. It is clear that 'the colourizing of cytogenetics' has occurred[11]. The next five years will be exciting as we apply the colourful technological innovations to diagnostic settings and research applications. To paraphrase a well known American author and would-be scientist 'One FISH, two FISH, red FISH, blue FISH, black FISH, blue FISH, old FISH, new FISH. Say, what a lot of FISH there are' (*Dr. Seuss*). ❏

1. Speicher, MR, *et al. Nature Genet.* **12**, 368-375 (1996).
2. Nederlof, PM, *et al. Cytometry* **10**, 20-27 (1989).
3. Nederlof, PM, *et al. Cytometry* **11**, 126-131 (1990).
4. Ried, T, *et al. Proc. Natl. Acad. Sci. USA* **89**, 1388-1392 (1992).
5. Dauwerse, JG, *et al. Hum. Molec. Genet.* **1**, 593-598 (1992).
6. Giam, X-Y, *et al. Genomics* **22**, 110-107 (1994).
7. Francke, U. *Cytogenet. Cell Genet.* **65**, 205-219 (1994).
8. Guan, X-Y, *et al. Nature Genet.* **12**, 10-11 (1996).
9. Romana, SP, *et al. Blood* **86**, 4263-4269 (1995).
10. Guan, X-Y, *et al. Nature Genet.* **8**, 155-161 (1994).
11. Ledbetter, DH. *Hum. Molec. Genet.* **1**, 297-299 (1992).
12. Trask, B. *Trends Genet.* **7**, 149-154 (1991).
13. Gray, JW, *et al. Curr. Opin. Biotech.* **3**, 623-631 (1992).
14. Le Beau, MM. in *Important Advances in Oncology.* (eds Devita VT, Hellman S, Rosenberg SA) 29-45 (J.P. Lippincott, Philadelphia, 1993).
15. Richkind, KE. *Applied Cytogenet.* **18**, 6-9 (1992).

* * * * * * * * * * * * * * * * * * * *

Questions:

1. What was one of the earliest limitations of using FISH?

2. Why was the t(12;21)(p13;q22) abnormality not cytogenetically detectable in childhood B cell acute lymphoblastic leukemia?

3. Why has FISH analysis not had the impact on clinical diagnosis that was originally envisioned when it was first introduced in the late 1980s?

Answers are at the end of the book.

Part Four

Gene Linkage
and
Chromosome Mapping

- 15 -

Stephen King, the well-known author of science fiction and horror novels, wrote a best-seller book in 1994 called Insomnia. *The main character of this book suffered from insomnia that progressed rapidly over a period of a few months. Stephen King based his character's affliction upon a rare inherited prion disease known as fatal familial insomnia (FFI). This autosomal dominant condition is characterized by severe sleep disruptions and with disturbances of the other bodily systems, including the autonomic, endocrine, and motor systems. Onset is usually in the fifth decade of life, and the disease duration is variable from 7 to 36 months. In Italy, an individual with FFI was found to be homozygous for the* PRNP *gene and demonstrated clinical and pathologic features typically found in heterozygotes.*

Fatal Familial Insomnia

by G. Rossi, *et al.*

Neurology, March 1998

Genetic, neuropathologic, and biochemical study of a patient from a new Italian kindred.

Article abstract — Fatal familial insomnia (FFI) is an inherited prion disease linked to a mutation at codon 178 of the *PRNP* gene that results in aspartic acid to asparagine substitution, in coupling phase with methionine at position 129. The disease is characterized clinically by insomnia with disturbances of the autonomic, endocrine, and motor systems and neuropathologically by selective degeneration of the thalamus. Phenotypic variability is well known and has been linked to homozygosity or heterozygosity at *PRNP* codon 129. We report the clinical, neuropathologic, and biochemical findings and genomic analysis of a patient with FFI from a new Italian kindred. Although homozygous for methionine at codon 129, this patient showed some clinical and pathologic features most commonly found in heterozygotes. *Neurology* 1998:50:688-692

Fatal familial insomnia (FFI) is an autosomal dominant disorder characterized by untreatable insomnia with disturbances of the autonomic, endocrine, and motor systems.[1,2] The mean age at onset of symptoms is 50 years, and the disease duration varies from 7 to 36 months.[2] The neuropathologic hallmarks are severe neuronal loss and astrogliosis of the thalamus, particularly the anterior ventral and mediodorsal nuclei. Cortical spongiosis and astrogliosis have been observed in patients with a protracted course.[1-3]

FFI is linked to a mutation at codon 178 of the prion protein (PrP) gene (*PRNP*) resulting in aspartic acid to asparagine substitution.[4-6] The same mutation is associated with a subtype of familial Creutzfeldt-Jakob disease (CJD[178]). However, these two conditions differ at polymorphic codon 129 of *PRNP* in that the mutant allele encodes methionine (Met) in FFI and valine (Val) in CJD.[7] The variant PRNP genotypes linked to FFI and CJD[178] result in the formation of protease-resistant PrP isoforms (PrP[res]) that differ as to the size of the protease-resistant fragments and to the pattern of glycosylation.[8]

In FFI, the Met/Val polymorphism at codon 129 also influences clinical presentation, disease duration, extension of neuropathologic changes, and brain regional distribution of PrP[res].[2,7,9-12] In a large series of cases, patients with short (7 to 8 months), intermediate (10 to 18 months), and long (25 to 33 months) disease duration were recognized. Patients with short disease duration were homozygous at codon 129, whereas patients with long disease duration were heterozygous; patients with intermediate disease duration were either homozygous or heterozygous.[9] Severe sleep and autonomic disturbances prevailed over the postural and motor abnormalities in homozygous patients, whereas the opposite occurred in heterozygotes. Differences were also observed neuropathologically. In fact, the homozygotes showed minimal changes in addition to the thalamic degeneration; PrP[res] was confined to the thalamus, striatum, and brainstem, whereas it was barely detectable in the neocortex. At variance, spongiosis, neuronal loss, and gliosis were common features in heterozygotes, involving diffusely the neocortex; PrP[res] was present in most gray structures, reaching the highest concentration in the cerebral cortex.[2,9]

We recently identified a new Italian family with FFI. In this study, we report the clinical, pathologic, and biochemical features of a patient from this kindred who was homozygous at codon 129 and had a disease of intermediate duration with early and severe motor abnormalities.

Methods. *Case history.* A 51-year-old woman presented with malaise, apathy, anorexia, and loss of libido, followed by insomnia, visual hallucinations, and myoclonic jerks involving the upper and the lower limbs. Sleep disturbances emerged from the clinical observation but were not documented by polysomnography. A few months later, the patient displayed disinhibited and aggressive behaviour and showed progressive impairment of short-term memory, learning and calculating ability, followed by global intellectual deterioration. Mild abnormalities of cerebellar function with bilateral dysmetria and unsteady gait became apparent. The myoclonic contractions were subcontinuous, particularly at the right arm. Serial EEG recordings documented a progressive slowing of background activity; in advanced stages of disease, long-lasting and diffuse sequences of pseudoperiodic diphasic or triphasic slow waves

recurring at 1.5-second intervals were present. The patient elapsed into coma and died 14 months after the onset of symptoms.

The proband's mother died at age 42 and grandmother died at age 58 years after a clinically similar neurologic disorder.

Neuropathologic study. The brain was removed 24 hours after death and divided midsagittally. The left half was fixed in 4% formaldehyde, whereas the frontal, temporal, and occipital poles of the right cerebral hemisphere and one sample of the right cerebellar hemisphere were fixed in Carnoy solution. The remaining parts of the right hemibrain were frozen at −80°C. Light microscopy examination was carried out on formalin- and Carnoy-fixed, paraplast-embedded sections of cerebrum, cerebellum, and brainstem. The sections were stained with hemotoxylineosin, Nissl, Bodian, Heidenhain-Woelcke, and thioflavine S or immunolabeled with a rabbit anti-serum to glial fibrillary acidic protein (GFAP, Dako, Copenhagen, Denmark; 1:600) and monoclonal antibodies to vimentin (Dako, 1:20) AND microtubule-associated protein (MAP)2 (Boehringer, Mannheim, Germany; 1:100). The immunoreactions were revealed by the peroxidase-antiperoxidase method and diaminobenzidine as a chromogen.

The distribution and extent of nerve cell loss in thalamic nuclei were evaluated by the squared graticule counting method.[13] Formalin-fixed, paraplast-embedded blocks of the left cerebral hemisphere including the whole thalamus were cut serially, and one 20-µm-thick section every 300 µm was stained with Nissl. For each thalamic nucleus (defined according to Jones' classification),[14] 10 random fields of $0.81 mm^2$ were scanned using a square-subdivided graticule in a x10 eyepiece, and neurons with a visible nucleus were counted. The results were compared with those obtained in an age-matched control brain that was processed and analyzed following the same procedure. The following scale of severity of nerve cell loss was applied: severe, >70%; moderate, 30 to 70%; mild, 10 to 30%; nonsignificant, <10%.

DNA analysis. Genomic DNA was extracted

from frozen brain tissue, and a 777-bp *PRNP* fragment was amplified by polymerase chain reaction (PCR) as described previously.[15] To investigate the codon 178 mutation and determine the codon 129 genotype of *PRNP*, the 777-bp PCR product was digested with restriction enzymes *Asp I* and *Mae II*, respectively. The restriction digests were then fractionated on 1.2% agarose gels and visualized with ethidium bromide.

Immunoblot analysis. Samples of cerebral neocortex and archicortex (middle frontal gyrus, inferior parietal lobule, middle temporal gyrus, lateral occipital gyrus, cingulate gyrus, parahippocampal gyrus, hippocampus), caudate nucleus, thalamus (mediodorsal and ventrolateral nuclei), hypothalamus, amygdala, midbrain, and cerebellum were homogenized in nine volumes of lysis buffer (100 mM NaCl, 10 mM EDTA, 0.5% Nonidet P-40, 0.5% sodium deoxycholate, 10 mM Tris, pH 7.4). After centrifugation at 3,000 *g* for 10 minutes, the protein concentration in the supernatant was determined by the bicinchoninic acid assay (Pierce, Rockford, IL). Samples equivalent to 200 µg protein were mixed with equal volumes of 2x Laemmli sample buffer and digested with 20 µg/mL proteinase K for 1 hour at 37°C. Proteolysis was terminated by the addition of phenylmethylsulphonyl fluoride (1 mM final concentration). Some samples were then subjected to deglycosylation by digestion for 2 hours with recombinant glycopeptide *N*-glycosidase F (New England Biolabs, Beverly, MA) under the conditions specified by the manufacturer. The samples were fractionated on 12.5% tricine-SDS-polyacrylamide minigels under reducing conditions, electrophoretically transferred to polyvinylidene difluoride membranes (Immobilon, Millipore, Bedford, MA), and probed with the monoclonal antibody 3F4 (1:50,000) that recognizes an epitope corresponding to residues 109 to 112 of human PrP. Samples of cerebral cortex from six patients with sporadic CJD were processed following the same procedure and used as positive control subjects. These patients were selected on the basis of the *PRNP* genotype (homozygosity for either Met or Val at codon 129) and the presence of type 1 or type 2 PrP[res] in brain homogenates.[16] Immunoreactive bands were visualized by enhanced chemiluminescence (Amersham, Buckinghamshire, UK) and quantified by densitometry using a Mitsubishi Video Copy Processor (Model P68E), a IV-530 Contour Synthesizer (FOR-A Co. Limited, France), and the Bio-Profil package of softwares (Vilber-Lourmat, France), with appropriate correction for background absorption.[9] Specificity of the reactions was checked by probing the membranes with 3F4 preabsorbed with a synthetic peptide homologous to residues 101 to 119 of human PrP.

Results. *Genetic analysis.* Digestion of a 777-bp *PRNP* fragment amplified from DNA by PCR with the restriction enzymes *Asp I* and *Mae II* showed that the patient carried a mutation at codon 178 of *PRNP* resulting in Asp to Asn substitution and was homozygous for Met at codon 129.

Neuropathology. The main neuropathologic feature was the degeneration of the thalamus. Severe neuronal loss (>70%) and astrogliosis involved most thalamic nuclei including the anterior ventral nucleus, ventral anterior nucleus, parvicellular population of the mediodorsal nucleus, ventral lateral nucleus, centromedian, lateral dorsal nucleus, lateral posterior nucleus, and pulvinar. Other sectors of the thalamus, including the intralaminar, reticular, limitans-suprageniculate and ventral posterior nuclei, and the medial geniculate body, were involved to a lesser degree, whereas the magnocellular population of the mediodorsal nucleus and the lateral geniculate body were not affected.

Nerve cell loss and gliosis were remarkable in medial septal nuclei and inferior olives, particularly in the dorsal region, whereas they were moderate in the basal nucleus of Meynert, locus coeruleus, dentate nucleus, and granule and Purkinje cell layers of the cerebellum.

The cerebral cortex was also involved in the disease process. Histologic examination showed moderate neuronal loss and gliosis in the cingulate gyrus, insula, and associative areas of frontal, parietal, and temporal lobes. Immunostaining with anti-MAP2 antibody confirmed the reduction of nerve cell bodies and revealed a striking decrease of immunoreactive dendritic processes. GFAP and vimentin immunostaining showed the presence of reactive astrocytes mainly localized in the deep layers of the cerebral cortex and in the subcortical white matter. Small foci of spongiosis were present in the anterior half of the cingulate gyrus, the medial aspect of superior frontal gyrus, and the parahippocampal gyrus. The white matter of centrum ovale, capsula interna, esterna and extrema, and cerebellum showed loss of myelinated fibers and diffuse astrogliosis.

Characteristics and distribution of PrP^{res}. Immunoblot analysis showed that many brain regions of the FFI patient contained two major PrP^{res} fragments of 28 and 26 kDa and a minor fragment of 19 kDa, *N*-deglycosylation resulted in a single species of 19 kDa, indicating that the 28- and 26-kDa peptides corresponded to PrP glycoforms. The electrophoretic mobility of PrP^{res} was distinct from that of type 1 PrP^{res} and identical to that of type 2 PrP^{res} from sporadic CJD, as specified previously.[9,14] However, both type 1 and type 2 PrP^{res} could be distinguished from the FFI peptides on the basis of the glycosylation pattern. In particular, the most abundant species was the high-molecular-weight glycoform in FFI and the low-molecular-weight glycoform in CJD; furthermore, the unglycosylated peptide was barely detectable in FFI, whereas it was well represented in CJD.

In the FFI patient, the highest levels of PrP^{res} were found in limbic areas, particularly in the cingulate gyrus, amygdala, and entorhinal cortex. Large amounts of the protein were also present in the hypothalamus and midbrain, whereas relatively low levels were detected in the caudate nucleus, thalamus, neocortex, and cerebellum (table).

Discussion. The patient described herein belongs to a new Italian family of FFI, apparently unrelated to the Italian kindreds recognized previously.[1,5] The characteristics of this case were consistent with those of FFI patients of intermediate duration reported by Parchi *et al.*[9] because the clinical course of the disease was 14 months and high levels of PrP^{res} were present in the cerebral cortex, particularly in limbic areas, in

addition to subcortical structures. On the other

Table: *Brain regional distribution of PrP^{res} in the FFI patient*

Brain region	PrP^{res}*
Cingulate gyrus	100
Amygdala	76
Entorhinal cortex	63
Midbrain	55
Hippocampus, CA1	47
Hypothalamus	46
Hippocampus, subiculum	31
Caudate nucleus	28
Thalamus	
Ventrolateral	23
Mediodorsal	18
Cortex	
Frontal	15
Occipital	14
Temporal	14
Parietal	12
Cerebellum	12
White matter	Not found

*Numbers indicate relative amounts of PrP^{res} and are referred to the highest value obtained from densitometric analysis, which was considered as 100%.

hand, our patient — homozygous at codon 129 — showed clinical features commonly found in heterozygotes, such as myoclonic jerks, cerebellar signs, and intellectual deterioration. Furthermore, she exhibited pseudoperiodic triphasic slow waves at the EEG in more advanced stages of the disease, and nerve cell loss, gliosis, and focal microspongiosis were detected in the cerebral cortex as neuropathologic examination. Accordingly, the clinical presentation and the neuropathologic findings in our patient expand the range of variability of FFI patients homozygous at codon 129 and point to weak correlation between genotype and the phenotypic characteristic in the group of patients with intermediate disease duration.

Thalamic degeneration and MAP2 immunohistochemistry deserve a comment. A morphometric analysis showed the thalamus of our patient more extensively degenerated than in the original cases.[1] In fact, severe nerve cell loss was not confined to the anterior ventral and dorsomedial nuclei but affected nuclear groups pertaining to the motor, association, intralaminar, and reticular systems in addition to the limbic system. Extensive thalamic degeneration has been already reported in FFI[3] and more recently confirmed in another patient from the original family,[17] arguing that the pathologic correlate of the dysfunction of circadian rhythms in FFI may be a diffuse degeneration of the thalamus. Even though the mediodorsal area, which is a critical connection point for structures involved in wake-sleep regulation, is invariably severely affected, the complex clinical picture of this disorder may result from the degeneration of other thalamic nuclei.[18]

A remarkable finding in our patient was a severe reduction in MAP2 immunoreactivity in the cerebral cortex, indicating a change in the epitope recognized by the MAP2 antibody or, most likely, a loss of dendrites in the cerebral cortex.[19] This finding may represent an anatomic correlate of cortical dysfunction and indicates that anti-MAP2 immunohistochemistry, which was not performed previously in FFI, may be a useful too to evaluate the involvement of the cerebral cortex in this disorder. ❑

References

1. Lugaresi E, Medori R, Montagna P, *et al*. Fatal familial insomnia and dysautonomia with selective degeneration of thalamic nuclei. *N Engl J Med* 1986; **315**::997-1003.
2. Gambetti P, Parchi P, Petersen RB, *et al*. Fatal familial insomnia and familial Creutzfeldt-Jakob disease: clinical, pathological and molecular features. *Brain Pathol* 1995; **5**:43-51.
3. Manetto V, Medori R, Cortelli P, *et al*. Fatal familial insomnia: clinical and pathologic study of five new cases. *Neurology* 1992; **42**:312-319.
4. Medori R, Tritschler HJ, LeBlanc A, *et al*. Fatal familial insomnia, a prion disease with a mutation at codon 178 of the prion protein gene. *N Engl J Med* 1992; **326**:444-449.
5. Medori R, Montagna P, Tritschler HJ, *et al*. Fatal familial insomnia: a second kindred with mutation of prion protein gene at codon 178. *Neurology* 1992; **42**:669-670.
6. Peterson RB, Tabaton M, Berg L, *et al*. Analysis of the prion protein gene in thalamic dementia. *Neurology* 1992; **42**:1859-1863.
7. Goldfarb LG, Petersen RB, Tabaton M, *et al*. Fatal familial insomnia and familial Creutzfeldt-Jakob disease: disease phenotype determined by a DNA polymorphism. *Science* 1992; **258**:806-808.
8. Monari L, Chen SG, Brown P, *et al*. Fatal familial insomnia and familial Creutzfeldt-Jakob disease: different prion proteins determined by a DNA polymorphism. *Proc Natl Acad Sci USA* 1994; **91**:2839-2842.
9. Parchi P, Castellani R, Cortelli P, *et al*. Regional distribution of protease-resistant prion protein in fatal familial insomnia. *Ann Neurol* 1995; **38**:21-29.
10. Reder AT, Mednick AS, Brown P, *et al*. Clinical and genetic studies of fatal familial insomnia. *Neurology* 1995; **45**:1068-1075.
11. Montagna P, Cortelli P, Tinuper P, *et al*. Fatal familial insomnia: a disease that emphasizes the role of the thalamus in the regulation of sleep and vegetative functions. In: Guillelminault C, Lugaresi E, Montagna P, Gambetti P, eds. Fatal familial insomnia: inherited prion diseases, sleep, and the thalamus. New York: Raven Press, 1994; 1-14.
12. Gambetti P, Medori R, Manetto V, *et al*. Fatal familial insomnia: a prion disease with distinctive histopathological and genotypic features. In: Guillelminault C, Lugaresi E, Montagna P, Gambetti P, eds. Fatal familial insomnia: inherited prion diseases, sleep, and the thalamus. New York: Raven Press, 1994; 15-22.
13. Aherne W. Some morphometric methods for the central nervous system. *J Neurol Sci* 1975; **24**:221-241.
14. Jones EG. The thalamus. New York: Plenum Press, 1985.
15. Tagliavini F, Prelli F, Porro M, *et al*. Amyloid fibrils in Gerstmann-Sträussler-Scheinker disease (Indiana and Swedish kindreds) express only PrP peptides encoded by the mutant allele. *Cell* 1994; **27**:695-703.
16. Parchi P, Castellani R, Capellari S, *et al*. Molecular basis of phenotypic variability in sporadic Creutzfeldt-Jakob disease. *Ann Neurol* 1996; **39**:767-778.
17. Macchi G, Rossi G, Abbamondi AL, *et al*. Diffuse thalamic degeneration in fatal familial insomnia: a morphometric study. *Brain Res* 1997; **77**:154-158.
18. Macchi G, Talamo e sistemi di vigilanza. In: Terzano MG, Parrino L, Smirne P, eds. Il sonno in Italia. Milano: Poletto Edizioni, 1995; 2-14.
19. Caceres A, Binder LI, Payne MR, *et al*. Differential subcellular localization of tubulin and the microtubule-associated protein MAP2 in brain tissue as revealed by immunocytochemistry with monoclonal hybridoma antibodies. *J Neuro-sci* 1984; **4**:394-410.

* * * * * * * * * * * * * * * * * * * *

Questions:

1. What does the mutant allele in PRNP code for in FFI and in Creutzfeldt-Jakob disease?

2. What features did the patient described in this study have that were seen more often in heterozytoes?

3. What is one hypothesis why sleep patterns are disrupted in individuals with FFI?

Answers are at the end of the book.

16

X-linked recessive dyskeratosis congenita is a rare bone-marrow failure disorder linked to Xq28. Autosomal dominant and recessive forms of the disease have been identified, although these are rarer than the X-linked form. The disease is characterized by the early manifestation of discolored skin, nail dystrophy, and the main cause of death (80%) is due to progressive bone-marrow failure. There is tremendous amount of variability in the age of onset, severity of bone marrow failure, and range of congenital abnormalities. There is also an increased susceptibility to malignancy. Fine-mapping of the gene responsible for dyskeratosis congenita began in 1978 and it took more than 18 years before a group of investigators were able to find the gene, DKC1, believed to be responsible for this condition. Now that the gene has been discovered, prenatal diagnosis can be offered to couples known to be at risk. However, it may be some time before patients can benefit from this discovery.

Dyskeratosis and Ribosomal Rebellion

by Lucio Luzzatto & Anastasios Karadimitris

Nature Genetics, May 19, 1998

Dyskeratosis congenita (DC) is a rare inherited X-linked disorder with a spectrum of clinical manifestations that includes — by definition — remarkable changes of the skin and its appendages. The appearance of a patient who has reached adulthood is quite characteristic: he will be a young man with little hair, almost no eyebrows, evanescent nails, teeth in a poor state, patches of discolouration on his skin, and sometimes tearful (due to abnormal lacrimal ducts). Unfortunately, the condition is more than skin-deep because on closer examination, dysplastic lesions may be discovered in his mouth or elsewhere in the gastro-intestinal tract, and eventually these may undergo malignant transformation. In addition, the patient may develop bone-marrow failure — it is partly for this reason that in the last few years the literature on DC has shifted conspicuously from dermatology journals to haematology journals. Diagnosis has necessarily depended on clinical findings, and is, of course, made easier when more than one male in the family is affected. In the past,

one would have supposed that more information about the metabolic or cellular abnormalities of affected tissues[1] would be required before the causative genetic lesion could be identified. It bears witness to the power of genomics that this was not necessary in the case of DC. On page 32 of this issue, Nina Heiss, Stuart Knight, Tom Vulliamy and colleagues report how, in one swoop, they have identified the gene that is mutated in DC (ref. 2).

X marks the spot

Fine-mapping of the DC gene took off in 1978, facilitated by a remarkable family3 in which the disorder segregated in apparent linkage and in *trans* (with a single recombinant) to a G6PD deficiency locus, which soon thereafter was mapped to X128 (ref. 4). Over the next 18 years, the critical region was narrowed to about 4 Mb (refs 5,6), to between the polymorphic markers GABRA3 and *DXS1108*. Heiss *et al.*[2] systematically tested all 28 available cDNA probes derived from this map interval and struck gold when one of these lit up an abnormal Southern-blot fragment derived from a DC patient (HO), who turned out to have a partial deletion within the 'causative' gene (*DKC1*). Point mutations were found in four other patients. The icing on the cake was the finding that, while DC is an orphan disease (that is, it is rare and has received little attention). *DKC1* is not an orphan gene. The cDNA sequence bears striking homology with a whole host of ESTs from a variety of sources, as well as with several 'full-length' nucleotide sequences from mammalian, nematode and micro-organism databases. It is clearly the orthologue of the yeast *CBF5P* and of the rat *NAP57*. Cbf5p, originally isolated as a centromere-binding protein[7], is essential for the processing of pre-rRNA (ref. 8), and is associated with the box H + ACA subset of small nucelolar RNAs (snoRNAs) that guide the site-specific conversion of uracil (U) to pseudo-uracil (Ψ) in rRNA (ref. 9). NAP57, together with the nucleolar protein Nopp140, is localized during interphase in the dense fibrillar component (DFC) of the nucleolus: at prometaphase it disperses to the cytoplasm, and at telophase it begins to re-enter nucleoli (without associating with pre-nucleolar bodies; ref. 10). Thus, although the precise function of the *DKC1* protein product (christened 'dyskerin') remains to be determined, it seems likely that it is a nucleolar protein and responsible for early steps in rRNA processing, perhaps by mediating pseudouridylation. As suggested for NAP57 (ref. 11), it may act as a kind of chaperone in the assembly of ribosomes and in their export from the nucleolus to the cytoplasm.

As is often the case when a gene responsible for a genetic disease is first isolated, several new questions arise. Some are general, some are specific to X-linked diseases, and some are specific to DC. A general questions is, of course, whether individual mutations are responsible for variation in clinical expression — a question that addresses the issue of genotype-phenotype correlation. While a mouse dyskerin knock-out may be informative and already in progress, human geneticists are used to the notion that perhaps a human 'knock-out' is already available in the clinic. Indeed, if the

deletion in patient HO causes complete loss of dyskerin activity, then dyskerin is clearly not necessary for general development or survival. Until the precise position of the deletion has been pinpointed, however, this remains uncertain. Heiss *et al.* seem to favour the view that the patient with the deletion, although having a more severe phenotype, may still retain some dyskerin function. As for missense mutations, it seems reasonable to expect that as more are identified, critical functional domains of dyskerin will be identified.

With respect to X-linkage, it is interesting to note that DC, although rare, has been observed in disparate populations[12]. It is tempting to speculate that many cases may result from relatively recent mutations and that the mutation spectrum will therefore be wide — this was predicted for X-linked genes by J.B.S. Haldane about half a century ago[13], and his prediction has been already validated by a wealth of data on haemophilia[14] and Duchenne muscular dystrophy[15] (the case of G6PD deficiency is different because many polymorphic alleles have been selected by malaria[16]). At the phenotypic level, women who are heterozygous for *DKC1* mutations show evidence of selection against the blood cells with the mutated gene on the active X chromosome[17]. Presumably, this also occurs in the skin and, in most cases, protects them from clinical expression of DC. There is, however, a documented case of female DC[18], and recently a well documented case of Wiskott-Aldrich syndrome in a young girl[19].

Dyskerin and ribosome activity

With respect to DC as a pathophysiological and clinical entity, many questions remain to be clarified. First, in view of the apparently ubiquitous expression of dyskerin, one wonders why the presenting manifestations are limited to certain tissues. One possibility is that there is some redundancy in the function of dyskerin and that its function becomes qualitatively or quantitatively critical in cells with a high turnover rate, such as those in the skin, in certain mucous membranes and in the bone marrow. Second, it has yet to be determined whether *DKC1* mutations act in a quantitative and/or qualitative fashion. On the one hand, a mutant dyskerin might become rate-limiting for pre-rRNA processing and therefore for ribosome biosynthesis. On the other hand, it could be that inadequately processed rRNA makes ribosomes functionally abnormal. We can probably rationalize the basic phenotype of DC through either or both of these mechanisms. Neither provides an immediate explanation for the apparently inarrestable progression of the disease, which includes gradual loss of skin derivatives and bone-marrow stem cells. These are features of premature ageing, although DC has never been classed among the premature ageing disorders. In this respect, an intriguing link between ageing and the nucleolus has been recently outlined[20]; dyskerin appears to be a nucleolar protein that, when dysfunctional, causes changes reminiscent of ageing-related conditions.

The nature of the link between the basic DC phenotype and the development of malignant tumours in DC patients remains obscure. At the human level, this is serious because — as is the case with several other cancer-susceptibility syndromes — it places the patient in double jeopardy: there is a certain phenotype caused by an inherited mutation and there is also the threat imposed by the uncertain stochastic nature of superimposed somatic mutations. We may soon be confronted with the question of how a defective ribosome chaperone can increase the risk of somatic mutations. This increase does not seem to be accounted for by an increase in cell turnover (in fact, cells appear to replicate at a slower rate, perhaps because their ribosomes are abnormal). Notably, a deficit of ribosomes causes the bobbed phenotype in *Drosophila* (the flies develop normally but are small; ref. 21). It may be that the abnormal ribosomes cause a new mutator phenotype through aberrant translation, in line with the error-catastrophe model of ageing[22] and consistent with the fact that mutations affecting ribosome activity are known to cause mis-coding in bacteria[23]. Once again, it may be some time before

patients benefit from this pivotal advance, although it will, of course, be possible to offer prenatal diagnosis to couples known to be at risk. In the meantime, they may provide biologists with new insights into ribosome function, as they would appear to have a ribosomapathy. ❏

References
1. Dokal, I. *et al. Blood* **80,** 3090-3096 (1992).
2. Heiss, N.S. *et al. Nature Genet.* **19,** 32-38 (1998).
3. Gutman, A., Frumkin, A., Adam, Al, Bloch-Stacher, N. & Rosenszain, L.A. *Arch. Dermatol.* **114,** 1667-1671 (1978).
4. Pai, G.S., Sprenkle, J.A., Do, T.T., Mareni, C.E. & Migeon, B.R. *Proc. Natl. Acad. Sci. USA* **77,** 2810-2813 (1980).
5. Connor, J.M. *et al. Hum. Genet.* **72,** 348-351 (1986).
6. Knight, S.W. *J. Med. Genet.* **33,** 993-995 (1996).
7. Jiang, W., Middleton, K., Yoon, H.J., Fouquet. C & Carbon, *J. Mol. Cell. Biol.* **13,** 4884-4893 (1993).
8. Cadwell, C., Yoon, H.-J., Zebarjadian, Y. & Carbon, *J. Mol. Cell. Biol.* **17,** 6175-6183 (1997).
9. Lafontaine, D.L.J., Bousquet-Antonelli, C. Henry, Y., Caizergues-Ferrer, M. & Tollervey, D. *Genes Dev.* **12,** 527-537 (1998).
10. Dundr, M. *et al. Chromosoma* **105,** 407-417 (1997).
11. Meier, U.T. & Blobel, G. *J. Cell Biol.* **127,** 1505-1514 (1994).
12. Sirinavin, C. & Trowbridge, A.A. *J. Med. Genet.* **13,** 339-354 (1975).
13. Haldane, J.B.S. *J. Genet.* **31,** 317-326 (1998)
14. Antonarakis, S.E. & Kazazsian, H.H., Jr. *Trends Genet.* **4,** 233-237 (1988).
15. Emery, A.E.H. *Duchenne Muscular Dystrophy.* (Oxford University Press, Oxford, 1988).
16. Luzzatto, L. & Mehta, A. in *The Metabolic and Molecular Bases of Inherited Disease,* (eds Scriver, C.R., Beaudet, A.L., Sly, W.S. & Valle, D.) 3367-3398 (McGraw-Hill, New York, 1995).
17. Vulliamy, T.J., Knight, S.W., Dokal, I. & Mason, P.J. *Blood* **90,** 2213-2216 (1997).
18. Sorrow, J.M. & Hitch, J.M. *Arch. Dermatol.* **88,** 340-347 (1963).
19. Parolini *et al. N. Eng. J. Med.* **338,** 291-295 (1998).
20. Guarente, L. *Genes Dev.* **11,** 2449-2455 (1997).
21. Boncinelli, E., Graziani, F., Polito, L., Malva, C. & Ritossa, F. *Cell Diff.* **1,** 133-142 (1972).
22. Orgei, L.E. *Nature* **243,** 441-445 (1973).
23. Gorini, L. *Nature New Biol.* **234,** 261-264 (1971).
24. Chen, E.Y. *et al. Hum. Mol. Genet.* **5,** 659-668 (1996).

* * * * * * * * * * * * * * * * * * * *

Questions:

1. What are the characteristic features of a male with dyskeratosis congenita?

2. What is the role of the DKC1 protein product?

3. How does the association of dyskeratosis congenita and the development of malignant tumors put affected individuals in "double jeopardy?"

Answers are at the end of the book.

17

Parkinson's disease is the second most common form of neurodegenerative disease after Alzheimer's, affecting 250,000-500,000 individuals in the United States. Characteristics of the disease include movement disorders composed of rigidity, resting tremors, and slowness in initiating and carrying out movement and difficulties remaining posture. Dysfunction and loss of neurons that produce the neurotransmitter dopamine in a region of the brain known as the substantia nigra are responsible for the symptoms associated with Parkinson's disease. Researchers in Japanese neurology have recently discovered a gene responsible for a rare autosomal recessive form of parkinsonism, AR-JP. The movement disorder, AR-JP, develops in teens or young adulthood and usually incapacitates patients by the 2nd to 3rd decade of life. Further studies will delineate whether mutations that cause AR-JP play a role in Parkinson's disease.

Mutations in the *parkin* Gene Cause Autosomal Recessive Juvenile Parkinsonism

by T. Kitada, *et al.*

Nature, April 9, 1998

Parkinson's disease is a common neurodegenerative disease with complex clinical features[1]. Autosomal recessive juvenile parkinsonism (AR-JP)[2,3] maps to the long arm of chromosome 6 (6q25.2q27) and is linked strongly to the markers D6S305 and D6S253 (ref. 4); the former is deleted in one Japanese AR-JP patient[5]. By positional cloning within this microdeletion, we have now isolated a complementary DNA clone of 2,960 base pairs with a 1,395-base-pair open reading frame, encoding a protein of 465 amino acids with moderate similarity to ubiquitin at the amino terminus and a RING-finger motif at the carboxy terminus. The gene spans more than 500 kilobases and has 12 exons, five of which (exons 3-7) are deleted in the patient. Four other AR-JP patients from three unrelated families have a deletion affecting exon 4 alone. A 4.5-kilobase transcript that is expressed in many human tissues but is abundant in the brain, including the substantia nigra, is shorter in brain tissue from one of the groups of exon-4-deleted patients. Mutations in the newly identified gene appear to be responsible for the pathogenesis of AR-JP, and we have therefore named the protein product 'Parkin.'

To investigate a small chromosomal deletion in a Japanese AR-JP patient, we first used screening by polymerase chain reaction (PCR) with a set of amplimers for *D6S305* to isolate two clones, KB761D4 and KB430C4 (approximate insert size 110 kilobases (kb) each) from a Keio human BAC (bacterial artificial chromosome) library[6]. Using exon trapping with these two BAC DNAs, we unexpectedly found only one putative exon (designated J-17) of 137 base pairs (bp). Amplification by PCR of the patient's DNA using two sets of amplimers, created from the J-17 sequence and its flanking regions, revealed that the patient has a deletion of J-17 in addition to one in *D6S305*.

We screened human fetal brain and skeletal muscle cDNA libraries with J-17 as an initial probe and then with the cDNAs, once isolated, as secondary probes. Seven cDNA clones were obtained and sequenced. The aligned sequence of 2,960 bp revealed an open reading frame of 1,395 bp (nucleotides 102-1,496), which encodes a protein of 465 amino acids with a relative molecular mass (M_r) 51,652. Four of seven cDNA clones lost a small sequence of 84 bp (nucleotides 636-719: amino-acid residues 179-206) in frame, implying that alternative splicing was involved.

The deduced amino-acid sequence of the newly identified protein showed moderate similarity to ubiquitin at the N terminus from Met1 to Lys76 (identities, 32%; positive, 62%). The lysine residue at position 48, which is essential for multi-ubiquitin chain formation[7], is conserved together with its flanking residues. Moreover, the remainder of the Parkin protein is rich in cysteine, and the C terminus has a motif similar to a RING-finger motif[8,9]: the sequence of Cys-X_2-Cys--X_9-Cys-X_1-His-X_2-[Cys-X_4-Cys]-X_4-Cys-X_2-Cys at the C terminus of Parkin matches the consensus sequence of the RING=finger apart from at the cysteine residues indicated in brackets. Parkin can therefore be considered as a member of the RING-finger family and/or as a new zinc-finger protein.

We next determined the genomic organization and exon/intron boundary sequences of *parkin*. As initially isolated, two clones contained only one putative exon sequence (J-17), so we also screened the Keio BAC library and a commercial source and obtained an additional 27 clones. This extraordinary number of clones implicated in the prospective gene was large, and we estimate *parkin* to be over 500 kb, with large introns (S.A. et al., manuscript in preparation). The sequence of genomic regions corresponding to the entire cDNA was determined by direct sequencing of selected clones, and the exon-intron organization and their boundary sequences were established (lodged under accession number AB009973 in the DDBJ database). The *parkin* gene consists of 12 exons, with exon 7 corresponding to the putative exon J-17 obtained by the initial exon trapping. Based on these data, we designed 14 sets of PCR primers to amplify each of 12 exons and determined the size of the expected PCR products.

PCR amplification of the members of family 1 revealed that patient IV-1 lacked five exons (exon 3-7), as confirmed by genomic Southern blot analysis. Thus, the microdeletion in patient IV-1 is intragenic, and the DNA for seven exons (exons 1, 2 and 8-12) is retained. We analysed two other patients from another unrelated family (family 2, patients II-1 and II-2) and found a deletion in exon 4 in these patients. This observation was confirmed by PCR analysis with reverse transcription (RT-PCR) of RNA extracted from the brain of patient II-2. The RT-PCR product is 633 bp for a healthy member but only 511 bp for the patient. Direct sequencing of the patient PCR product indicated that exon 3 is next to exon 5 with exon 4 missing. This exon-4 deletion was found in two additional AR-JP patients from two other unrelated families. Thus, the deletion mutation was found in five Japanese AR-JP patients from four independent families. We conclude that the gene encoding the Parkin protein must be responsible for the pathogenesis of AR-JP.

Northern blot analysis, using poly(A)+RNA with J-17 (exon 7) as a probe, showed that a 4.5-kb transcript is expressed in many human tissues, including brain, heart, testis, and skeletal muscle. In the brain, this gene was expressed in various regions, including substantia nigra.

The characteristic clinical symptoms of AR-JP include a susceptibility to dopa-induced dyskinesia and motor fluctuation, in addition to levodopa-responsive parkinsonism[23]. Pathologically, it is characterized by the loss of nigral and locus coeruleus neurons without Lewy body formation[10]. Parkinson's disease is probably initiated by a combination of genetic predisposition and environmental factors[11,12], which eventually induce mitochondrial respiratory failure and oxidative stress in nigral neurons[12-14]. A mutation in α-synuclein, a component of the Lewy bodies[15], is linked to the autosomal dominant form of Parkinson's disease[16]. We have described a second Parkinson's disease gene which encodes a new protein, 'Parkin,' linked to early onset disease that causes neurodegeneration in the substantia nigra without Lewy body formation.

Parkin is similar to the ubiquitin family of proteins, which are involved in the pathogenesis of several neurodegenerative diseases and are a component of the paired helical filaments in Alzheimer's disease[17-19] and of Lewy bodies in Parkinson's disease[20-22]. In the paired helical filaments of Alzheimer's disease, ubiquitin is conjugated with tau, but the majority of paired helical filaments remain as mono-ubiquitinated forms, which are not recognized by proteasomes and hence not processed[18]. Amyloid β protein inhibits the ubiquitin-dependent proteasome pathway *in vitro*[19]. On the other hand, Lewy bodies contain abundant multi-ubiquitin chains[22]. Incomplete degradation of Lewy-body-associated proteins may cause deposition of multi-ubiquitinated proteins in Lewy bodies[22], so ubiquitin or ubiquitin-like proteins may be important in the pathogenesis of Parkinson's disease.

Parkin may function similarly to ubiquitin family members, and its defect in AR-JP may interfere with the ubiquitin-mediated proteolytic pathway leading to the death of nigral neurons[7,23-25]. Note, however, that no Lewy body formation occurs in AR-JP, in which the *parkin* gene appears to be defective. The ubiquitin portions of several ubiquitin-conjugated proteins also act as chaperones for the remaining portions[26], so the ubiquitin-like portion of Parkin may be involved in its functional maturation. Then the C-terminal portion of Parkin containing the RING-finger motif may function as a new zinc-finger protein in the control of cell growth, differentiation, and development[8]. Further investigation is necessary to establish the exact physiological function of Parkin and how the parkin gene defect induces selective degeneration of nigral neurons in AR-JP without Lewy body formation. ❏

References

1. Parkinson, J. *An Essay on the Shaking Palsy* (Whittingham and Rowland, London, 1817).
2. Yamamura, Y., Arihiro, K., Kohriyama, T. & Nakamura, S. Early-onset parkinsonism with diurnal fluctuation-clinical and pathological studies. *Clin. Neurol. (Tokyo)* **33**, 491-496 (1993).
3. Ishikawa, A. & Tsuji, S. Clinical analysis of 17 patients in 12 Japanese families with autosomal recessive type juvenile parkinsonism. *Neurology* **47**, 160-166 (1996).
4. Matsumine, H. *et al.* Localization of a gene for autosomal recessive form of juvenile parkinsonism (AR-JP) to chromosome 6q25.2-27. *Am. J. Hum. Genet.* **60**, 588-596 (1997).
5. Matsumine, H. *et al.* Evidence for a microdeletion of D6S305 in a family of autosomal recessive juvenile parkinsonism (AR-JP). *Genomics* (in the press).
6. Asakawa, S. *et al.* Human BAC library: Construction and rapid screening. *Gene* **191**, 69-79 (1997).
7. Finley, D. & Chau, V. Ubiquitination. *Annu. Rev. Biol.* **7**, 25-69 (1991).
8. Saurin, A.J., Borden, K.L.B., Boddy, M.N. & Freemont, P.S. Does this have a familiar RING? *Trends Biochem. Sci.* **21**, 208-214 (1996).
9. Aasland, R., Gibson, T.J. & Stewart, A.F. The PHD finger: implications for chromatin-mediated transcriptional regulation. *Trends Biochem. Sci.* **20**, 56-59 (1996).
10. Takahashi, H. *et al.* Familial juvenile parkinsonism: clinical and pathologic study in a family. *Neurology* **44**, 437-441 (1994).
11. Calne, D.B. & Langston, J.W. Aetiology of Parkinson's disease. *Lancet ii*, 1456-1459 (1983).
12. Jenner, P., Schapira, A.H.V. & Marsden, C.D. New insights into the cause of Parkinson's disease. *Neurology* **42**, 2241-2250 (1992).
13. Schapira, A.H.V. *et al.* Mitochondrial complex I deficiency in Parkinson's disease. *J. Neurochem.* **54**, 824-827 (1990).
14. Mizzuno, Y. *et al.* Role of mitochondria in the etiology and pathogenesis of Parkinson's disease. *Biochim. Biophys. Acta* **1271**, 265-274 (1995).
15. Spillantini, M.G. *et al.* α-Synuclein in Lewy bodies. *Nature* **388**, 839-840 (1997).
16. Polymeropoulos, M.H. *et al.* Mutation in α-synuclein gene identified in families with Parkinson's disease. *Science* **276**, 2045-2047 (1997).
17. Mori, H., Kondo, J. & Ibara, Y. Ubiquitin is a component of paired helical filaments in Alzheimer's disease. *Science* **235**, 1641-1644 (1987).
18. Morishima-Kawashima, M. *et al.* Ubiquitin is conjugated with amino-terminally processed tau in paired helical filaments. *Neuron* **10**, 1151-1160 (1993).
19. Gregori, L., Fuchs, C., Figueiredo-Pereira, M.E., Nostrand, W.E.V. & Goldgaber, D. Amyloid β-protein inhibits ubiquitin-dependent protein degradation *in vitro*. *J. Biol. Chem.* **270**, 19702-19708 (1995).
20. Love, S., Saitoh, T., Quijada, S., Cole, G.M. & Terry, R.D. Alz-50, ubiquitin and Tau immunoreactivity of neurofibrillary tangles, Pick bodies and Lewy bodies. *J. Neuropathol. Exp. Neurol.* **47**, 393-405 (1988).
21. Galloway, P.G., Mulvihill, O. & Perry, G. Filaments of

Lewy bodies contain insoluble cytoskeletal elements. *Am. J. Pathol.* **140,** 809-822 (1922).

22. Iwatsubo, T. *et al.* Purification and characterization of Lewy bodies from the brains of patients with diffuse Lewy body disease. *Am. J. Pathol.* **148,** 1517-1529 (1996).
23. Ciechanover, A. The ubiquitin-proteasome proteolytic pathway. *Cell* **79,** 13-21 (1994).
24. Hochstrasser, M. Ubiquitin, proteasomes, and the regulation of intracellular protein degradation. *Curr. Opin. Cell Biol.* **7,** 215-223 (1995).
25. Gregori, L., Poosch, M.S., Cousins, G. & Chau, V. A uniform isopeptide-linked multiubiquitin chain is sufficient to target substrate for degradation in ubiquitin-mediated proteolysis. *J. Biol. Chem.* **265,** 8354-8357 (1990).
26. Finley, D., Bartel, B. & Varshavsky, A. The tails of ubiquitin precursors are ribosomal proteins whose fusion to ubiquitin facilitates ribosome biogenesis. *Nature* **338,** 394-401 (1989).
27. Sambrook, J., Fritsch, E.F. & Maniatis, T. *Molecular Cloning: A Laboratory Manual* (Cold Spring Harbor Laboratory Press, Cold Spring Harbor, New York, 1989).

* * * * * * * * * * * * * * * * * * * *

Questions:

1. Name the major differences between AR-JP and Parkinson's disease.

2. True or False? Both Parkinson's disease and AR-JP are associated with a mutation in α-synuclein.

3. What two critical residues does *parkin* lack that are important for ubiquitination?

Answers are at the end of the book.

1. AP-JP   Park
   - autosmal recessive   - complex inheritence
   - no lewy bodies   - lewy bodies
   - mutations are deletions - missense mutations

3.

- 18 -

Hair is an important and characteristic feature of mammals, and various pathological conditions, including infectious diseases, hormonal disturbances and hereditary disorders, can lead to loss of scalp and body hair. There are several forms of hereditary baldness — known as alopecia — transmitted as autosomal, recessive, or X-linked traits, but examples of sex-influenced inheritance have also been described. Severe, premature baldness is considered to be inherited in an autosomal dominant manner, with gene expression limited to the male, unless it is present in the homozygous state. Premature balding is a feature of myotonic dystrophy, a variety of ectodermal dysplasias and specific hair disorders. However, with the exceptions of the cases, a clear genetic origin for alopecia has yet to be discovered. Researchers from Columbia University in New York have mapped the first genetic defect causing human hereditary hair loss to chromosome 8p12.

Alopecia Universalis Associated with a Mutation in the Human *hairless* Gene

by Wasim Ahmad, *et al.*

Science, January 30, 1998

There are several forms of hereditary human hair loss, known collectively as alopecias, the molecular bases of which are entirely unknown. A kindred with a rare, recessively inherited type of alopecia universalis was used to search for a locus by homozygosity mapping, and linkage was established in a 6-centimorgan interval on chromosome 8p12 (the logarithm of the odds favoring linkage score was 6.19). The human homolog of a murine gene, **hairless,** *was localized in this interval by radiation hybrid mapping, and a missense mutation was found in affected individuals. Human* **hairless** *encodes a putative single zinc finger transcription factor protein with restricted expression in the brain and skin.*

The human hair follicle is a dynamic structure that generates hair through a complex and exquisitely regulated cycle of growth and remodeling (1). Despite the extensive descriptive understanding of the hair cycle, currently, very little is known about the molecular control of the signals that regulate progression through the hair cycle, although it is clear that at least some potentially influential regulatory molecules may play a role (1). For example, a knock-out mouse with targeted ablation of the gene encoding the fibroblast growth factor 5 (FGF5) provided evidence that FGF5 is an inhibitor of hair elongation, and the mouse had an increase in hair length due to an increase in the time that follicles remain in anagen. The FGF5 gene was also deleted in the naturally occurring mouse model, *angora* (2). Another member of the FGF family, FGF7 or keratinocyte growth factor, was disrupted by gene targeting, and the resultant mouse had hair with a greasy, matted appearance, similar in phenotype to the rough mouse (3). A transgenic mouse was engineered that disrupted the spatial and temporal expression of the gene encoding the lymphoid enhancer factor 1, a transcription factor that binds to the promoter region of several published hair keratin promoters. Disruption of this potential master regulator of hair keratin transcription resulted in defects in the positioning and angling of the hair follicles (4). More recently, a mutation in a structural protein mouse desmoglein 3 (encoded by the gene *dsg3*), was found to be the underlying mutation in the naturally occurring mouse phenotype, *balding* (5). Finally the *nud* mouse phenotype, characterized by hairlessness and athymia, was found to be the result of mutations in the winged-helix *nud* (*whn*) gene, a member of the winged-helix class of transcription factors (6). In addition to the complexity of the signaling pathways in sheep, there are over 100 distinct structural proteins synthesized by the hair cortex and cuticle cells that produce the keratinized structure of a wool fiber (1). Despite these examples of recent progress in animal models, we have only begun to understand the control and molecular complexity of the hair follicle and its cyclic progression in humans.

There are several forms of hereditary human hair loss, known collectively as alopecias, which may represent a dysregulation of the cycle of hair growth and remodeling (1), yet the molecular basis of the alopecia has remained largely unexplored (7). The most common form of hair loss, known as androgenetic alopecia (male pattern baldness), is believed by some to affect ~80% of the population (7). Alopecia areata is a common dermatologic disease affecting about 2.5 million individuals in the United States alone, which causes round, patchy hair loss on the scalp (7). Alopecia areata can progress to involve hair loss from the entire scalp; this condition is referred to an alopecia totalis. Alopecia universalis (AU) is the term for the most extreme example of disease progression, which results in the complete absence of scalp and body hair (7). Although an autoimmune patho mechanism for alopecia areata has been suggested, the precise etiology is unknown, and no autoantigen or causative gene has been identified (8). The inheritance patterns of these forms of alopecia are also unclear, although a polygenic

model with variability in penetrance and expressivity would appear most plausible, perhaps modulated by superimposed hormonal or immune factors.

In an effort to understand the molecular basis of a simple, recessively inherited form of AU (Online Mendelian Inheritance in Man accession number 203655) with no evidence of a confounding autoimmune component, we studied a large Pakistani kindred with AU segregating as a single Mendelian abnormality without associated ectodermal defects and containing four affected males and seven affected females. The affected individuals were in good general health, with no evidence for immune system dysfunction or unusual susceptibility to skin tumors. At birth, the hair usually appears normal on the scalp but never regrows after a ritual shaving, usually performed a week after birth. A skin biopsy from the scalp of an affected person revealed very few hair follicles, which were dilated and without hairs, and the absence of an inflammatory infiltrate. Affected individuals are born completely devoid of eyebrows and eyelashes and never develop axillary and pubic hair. The pedigree is strongly suggestive of autosomal recessive inheritance, and the large number of consanguineous loops account for all affected persons being homozygous for the abnormal allele.

To identify the alopecia locus segregating in this family, we initiated a genome-wide search for linkage by homozygosity mapping (9). During the initial screening, DNA samples from four affected individuals were genotypes with 386 highly polymorphic microsatellite markers spaced at 10-centimorgan (cM) intervals (10). In the course of this screen, 13 genomic regions were found to be homozygous for three to four affected individuals; each of these genomic regions were tested further in 32 additional family members, and 12 of the regions were excluded. One marker, D8S136 on chromosome 8p12, was found to be homozygous in all seven living affected individuals. Further analysis with markers from this region resulted in the identification of homozygosity in all affected individuals for the markers D8S1786 and D8S298 (11). Allele patterns obtained with the markers

D8S136 and D8S1786 indicated that these two markers are placed very close to each other on chromosome 8p12. A maximum two-point logarithm of the odds ratio for linkage (lod) score of 6.19 at zero recombination was achieved with the marker D8S298 by means of the FASTLINK 3.0 package (12), indicating that the alopecia gene in this family maps to chromosome 8p12. Recombinant haplotypes observed in individuals place the alopecia locus within a 6-cM interval between the distal and proximal marker D8S258 and D8S1739, respectively, with no obvious candidate gene in this interval.

In an independent line of investigation, we had developed an interest in the *hairless* mouse (13) as a potential model for inherited human alopecias and had begun to clone the human homolog of *hairless* with polymerase chain reaction (PCR) primers based on the available murine cDNA sequence (Gene Bank accession number Z32675) (13). We reverse transcriptase (RT)-PCR amplified a segment corresponding to exons 13 to 18 in the murine sequence using human skin fibroblast mRNA as template (14) and subsequently delineated the entire coding sequence of human *hairless*, which consists of 1189 amino acids. The expression pattern of human *hairless* is consistent with that observed in mouse (13) and rat (15), with substantial expression in the brain and skin and trace expression elsewhere (16). The human and mouse amino acid sequences are 84% homologous and 80% identical, and the human and rat sequences are 83% homologous and 78% identical. The murine *hairless* gene resides on mouse chromosome 14 (13), which shares synteny with human chromosomes 8p and 14q, among others (17). To determine the precise chromosomal localization of the human homolog of *hairless*, we used radiation hybrid mapping (18) with the GeneBridge 4 panel consisting of 93 radiation-induced human-hamster cell hybrids (Research Genetics), which placed the human homolog of the mouse *hairless* gene on chromosome 8p, between the two polymorphic markers D8S280 and D8S278, spanning a 19-cM region. The 6-cM candidate region obtained for the AU gene by linkage analysis with flanking markers D8S258

and D8S1739 lies between markers D8S280 and D8S278 on the basis of the Genome Data Base (17), the Center for Medical Genetics database (19), and the radiation hybrid map constructed by the Human Genome Mapping Center at Stanford University (20). On the basis of this genomic colocalization, the human *hairless* gene became a major candidate gene responsible for AU in this family, and the search for a mutation was initiated.

Direct sequence analysis (21) of exon 15 (amino acids 993 to 1032) revealed a homozygous A-to-G transition in all affected individuals, which was present in the heterozygous state in obligate carriers within the family and not found in unaffected family members. The A-to-G transition occurred at the first base of a threonine (T) residue at position 1022 (ACA), leading to a missense mutation and converting the threonine residue to an alanine (A) residue (GCA), and was designated T1022A. The mutation created a new cleavage site for the restriction endonuclease Hga I (GACGC), which was used to confirm the presence of the mutation in genomic DNA, in addition to direct sequencing (21). To verify that the missense mutation was not a normal polymorphic variant, we screened for the mutation by a combination of heteroduplex analysis (22), direct sequencing, and restriction digestion in a control population consisting of 142 unrelated, unaffected individuals, 87 of whom were of Pakistani origin. No evidence for the mutant allele was found in these individuals.

The *hairless* mouse, *hr/hr*, arose from spontaneous integration of an endogenous murine leukemia provirus into intron 6 of the *hairless* gene (23), resulting in aberrant splicing and only about 5% normal mRNA transcripts present in *hr/hr* mice (13). The protein encoded by the human, mouse, and rat *hairless* genes contains a single zinc finger domain with a novel and conserved six-cysteine motif and is therefore thought to function as a transcription factor (13, 15), with structural homology to the GATA family (24) and to TSGA, a protein encoded by a gene expressed in rat testis (25). In addition to the total body hair loss that bears striking resemblance to AU, the *hr/hr* mouse exhibits a number of phenotypic effects not observed in the AU family, including defective differentiation of thymocytes (26), as well as a unique sensitivity to ultraviolet radiation and chemically induced skin tumors (27). Similar to previous studies in mouse and rat, human *hairless* was substantially expressed in fibroblasts from hair-bearing skin and most highly expressed in brain, where its importance remains unknown (15). Evidence in support of a role for multifunctional transcription factors involved in hair loss can be drawn from genetic studies of the *whn* gene in *nude* mice, in which mutations in a winged-helix transcription factor result in the absence of hair and athymia (6). Molecular evolutionary studies of *whn* have shown that a homolog is present in the puffer fish, *Fugu rubripes*, which has a thymus, but not hair, therefore suggesting that its role in hair keratinization may represent a new function for *whn* in mammals (6). A COOH-terminal activation domain was identified in *whn* by comparative genome analysis. The essential function of this domain could be obliterated by site-directed mutagenesis of acidic residues to alanine, analogous to the missense mutation we describe in the AU family. It is possible that in humans, the AU mutation disrupts a similar potential activation domain within *hairless* with restricted specificity in the skin, whereas the *hr/hr* mouse displays a more pleiotropic defect because of the near absence of *hairless* mRNA and protein (6, 28). We anticipate that further studies into the biology of human hairless and its transactivation targets may illuminate potential therapeutic opportunities. ❑

References and Notes

1. M.H. Hardy, *Trends Genet.* **8**, 159 (1992); T.A. Rosenquist and G.R. Martin, *Dev. Dyn.* **205**, 379 (1996).
2. J.M. Hebert, T. Rosenquist, J. Gotz, G.R. Martin, *Cell* **78**, 1017 (1994).
3. L. Guo, L. Degenstein, E. Fuchs, *Genes Dev.* **10**, 165 (1996).
4. P. Zhou, C. Byrne, J. Jacobs, E. Fuchs, *ibid.* **9**, 700 (1995).
5. P.J. Koch *et al., J. Cell Biol.* **137**, 1091 (1997).
6. M. Nehls, D. Pfeifer, M. Schorpp, H. Hedrich, T. Boehm, *Nature* **372**, 103 (1994); J. Segre, J. L. Nemhauser, B.A. Taylor, J.H. Nadeau, E.S. Lander, *Genomics* **28**, 549 (1995): K. Schüddenkopf, M. Schorpp, T. Boehm, *Proc. Natl. Acad. Sci. U.S.A.* **93**, 9661 (1996).
7. A. Rook and K. Dawber, *Diseases of the Hair and Scalp* (Blackwell, Oxford, UK. ed. 2, 1991), p. 136; W.F. Bergfield, *Am. J. Med.* **98**, 95S (1995).
8. H.K. Muller *et al., Br. J. Dermatol.* **102**, 609 (1980).
9. V.C. Sheffield, D.Y. Nishimura, E.M. Stone, *Curr. Opin. Genet. Dev.* **5**, 335 (1995).
10. Blood samples were collected from 36 members of the AU family, according to local informed consent procedures. DNA was isolated according to standard techniques (29). Fluorescent automated genotyping for the genome-wide linkage search was carried out at as a service by Research Genetics, with 386 markers covering the genome at about 10-cM intervals.
11. Refined and more extensive screening all regions showing homozygosity in affected and unaffected family members was carried out with primers obtained from Research Genetics or in the Genome Data Base (17). Analysis of microsatellite markers consisted of end-labeling one primer with [γ-^{33}P]deoxyadenosine triphosphate; a PCR reaction consisting of 7 min at 95°C, followed by 27 cycles of 1 min at 95°C, 1 min at 55°C, and 1 min at 72°C; and electrophoresis in a 6% polyacrylamide gel (Sequa-gel; Action Scientific, Atlanta, GA). Microsatellite markers were visualized by exposure of the gel to autoradiography, and genotypes were assigned by visual inspection.
12. Statistical calculations for linkage analysis were carried out with the computer program FASTLINK version 3.0P [A.A. Schaffer, *Hum. Hered.* **46**, 226 (1996)], which enables all inbreeding loops in the family to be retained and has the capability for two-point analysis. Autosomal recessive with complete penetrance was assumed with a disease allele frequency of 0.0001. The lod scores were calculated with equal allele frequencies; however, results did not change when the frequency of the marker allele in association with the disease allele was set as high as 0.9. Multipoint analysis was not possible because of the large number of inbreeding loops and the complexity of the pedigree.
13. H.C. Brooke, *J. Hered.* **15**, 173 (1924); M.B. Cachon-Gonzalez *et al., Proc. Natl. Acad. Sci. U.S.A.* **91**, 7717 (1994).
14. For RT-PCR of human hairless cDNA sequences, total RNA was extracted from cultured skin fibroblasts from hair-bearing skin from a control individual according to standard methods (29). We reverse transcribed human *hairless* mRNAs with mouse mammary leukemia virus RT (Gibco-BRL), using an oligo-deoxyribosyithymine primer (Pharmacia). PCR was carried out with the following primers, constructed on the basis of the mouse hairless sequence (GenBank accession number Z32675): 5'-TGAGGGCTCTGTCCTCCTGC-3' (sense) and 5'-GCTGGCTCCCTGGTGGTAGA-3' (antisense). PCR conditions were 5 min at 95°C, followed by 35

cycles of 1 min at 95°C, 1 min at 55°C, and 1 min at 72°C, with AmpliTaq Gold DNA polymerase (Perkin-Elmer). After direct sequencing of the human cDNA, exon-based primers were designed and used to amplify genomic DNA as template and directly sequenced with the ABI 310 Automated Sequencer. The intron-exon borders were determined by comparison of the cDNA and the genomic DNA sequences at both the 5' donor and 3' acceptor splice junctions. The human *hairless* sequence has been deposited in GenBank (accession number AFO39196).

15. C.C. Thompson, *J. Neurosci.* **16**, 7832 (1996).
16. The human multiple tissue Northern blot containing 2 μg of polyadenyiated [poly(A)$^+$] mRNA from eight tissues was obtained from Clontech (Palo Alto, CA) and hybridized according to the manufacturers' recommendations with a random primed radiolabeled probe containing exons 13 to 18 of human *hairless*, generated as described in (14) and hybridized with ExpressHyb Solution (Clontech). Poly(A)$^+$ mRNA was extracted from cultured skin fibroblasts from hair-bearing skin from a control individual according to standard methods (29), and a Northern blot containing 2 μg of poly(A)$^+$ mRNA was hybridized with the same probe under identical conditions.
17. See the Genome Data Base at www.bis.med.jhmi.edu
18. A segment of human *hairless* intron 13 was PCR-amplified and used for radiation hybrid mapping with the G3 panel by Research Genetics. Primers were as follows: 5'-TATGTCACCAAGGGCCAGCC-3' (sense) and 5'-TCAGGGTAGGGGGTCATGCC-3' (antisense). PCR conditions were 5 min at 95°C, followed by 35 cycles of 1 min at 95°C, 1 min at 55°C, and 1 min at 72°C, with AmpliTaq Gold DNA polymerase (Perkin-Elmer). PCR primers specifically amplified human *hairless* and did not cross-hybridize with the hamster DNA used in the radiation hybrid panel.
19. See www.marshmed.org/genetics/maps/ss-maps.8ss.txt
20. See www.shgc.stanford.edu
21. Primers for specific amplification of exon 14 were placed in the flanking introns: 5'-AGTGCCAGGATT ACAGGCGT-3' (sense, intron 15) and 5'-CTGAGG AGGAAAGAGCGCTC-3' (sense, intron 16). PCP fragments were purified on AGTC Centriflex columns (Edge BioSystems, Gaithersburg, MD) and sequenced directly with POP-6 polymer on an ABI Prism 310 Automated Sequencer (Perkin-Elmer). The mutation was verified by restriction endonuclease digestion with Hga 1, according to the manufacturer's specifications (New England Biolabs).
22. A. Ganguly, M.J. Rock, D.J. Prockop, *Proc. Natl. Acad. Sci. U.S.A.* **90**, 10325 (1993).
23. J.P. Stoye, S. Fenner, G.E. Greenoak, J.M. Moral, **Cell 54**, 383 (1988).
24. R.J. Arceci, A.A.J. King, M.D. Simon, S.H. Orkin, D.B. Wilson, *Mol. Cell. Biol.* **13**, 2235 (1993).
25. C. Höög, M. Schalling, E. Grunder-Brundell, B. Danehoit, *Mol. Reprod. Dev.* **30**, 173 (1991).
26. P.J. Morrissey, D.R. Parkinson, R.S. Schwartz, S.D. Waksal, *J. Immunol.* **125**, 1558 (1980).
27. C.H. Gallagher, F.R.C. Path, P.J. Canfield, G.E. Greenoak, V.E. Reeve, *J. Invest. Dermatol.* **83**, 169 (1984).
28. G.L. Semenza, *Hum. Mutat.* **3**, 180 (1994); D.S. Latchman, *N. Engl. J. Med.* **334**, 28 (1996); D. Engelkamp and V. van Heyningen, *Curr. Opin. Genet. Dev.* **6**, 334 (1996).
29. J. Sambrook, E.G. Fritsch, T. Maniatis, *Molecular Cloning, A Laboratory Manual,* (Cold Spring Harbor Laboratory Press, Cold Spring Harbor, NY, ed. 2, 1989).
30. We sincerely thank the family members for their participation in this work; S. Malik and S.M. Zaidi for their assistance during our stay in Chakwai, B.J. Longley for expert dermatopathological advice, and M. Grossman and P. Schneiderman for their infinite patience. Informed consent for the publication of the photographs as well as the pedigree was obtained through personal visits to the family members. Supported in part by grants from Quaid-i-Azam University, Islamabad, Pakistan (M.A.), the National Alopecia Areata Foundation (A.M.C.), NIH-National Human Genome Research Institute HG-00008 (J.O.), and NIH-National Institute of Arthritis and Musculoskeletal and Skin Diseases Research Center P30AR44535 (M.P., J.O., and A.M.C.).

* * * * * * * * * * * * * * * * * * * *

Questions:

1. What role does the gene encoding fibroblast growth factor 5 play in mice?

2. What is the most common form of baldness and what percent of the population is affected by this form?

3. What was the genotype of the affected individuals in the Pakistani pedigree?

4. What method did the researchers use to identify the gene responsible for alopecia universalis?

Answers are at the end of the book.

Part Five

Regulation
of
Gene Expression

- 19 -

Gene therapy has been studied and practiced for several years. However, the effectiveness of this treatment approach has not been overwhelming. The most dramatic example was a four-year-old girl who was treated by gene therapy because she was suffering from severe combined immunodeficiency syndrome (SCIDS), or the lethal "bubble boy" disorder. One of the greatest challenges that scientists face in using this technique is getting the healthy gene to the target area. Multiple barriers are naturally present in the human, namely the immune system and the blood-brain barrier in the brain. Researchers at S. Louis University Health Sciences Center appear to have overcome the obstacles by using an experimental treatment called GLI-328 that may revolutionize the treatment for a deadly form of brain cancer.

A Head Full of Hope

by Jeff Goldberg

Discover, April 1998

To attack a terrifying form of brain tumor, surgeons are adding a tiny new tool to their kit: a genetically tweaked virus, designed to mark cancer cells for death.

Betty Perr's strange symptoms had begun just two weeks earlier, when the 67-year-old St. Louis resident thought she heard two of her grandchildren, Josephine, then 5, and Alexandria, 4, talking and giggling outside her house. The children lived just down the road with her daughter and son-in-law, Miriam and Robert Butler, and often dropped in to see their "Mama." But when she opened the door, they weren't there. That weekend, she heard the children's voices again — only this time she and her husband, Hugo, were 120 miles away at a wedding in Jefferson City. On the way home and over the following days, she complained about hearing a turkey gobbling, a dog barking, and the ghostly sound of a child's footsteps running through a house where she worked every week as a housecleaner.

After this last incident, Miriam Butler and her sister, Kathy Kramer, both nurses, urged their mother to see a doctor. A few days later Perr was admitted to the St. Louis University Health Sciences Center, where Miriam worked. A magnetic resonance imaging (MRI) scan revealed the cause of the phantom voices and noises to be a tumor in the right temporal lobe of her brain, in an area just above her ear. The location of the tumor, in a part of the brain thought to be involved in our ability to remember voices, songs, and other familiar sounds, accounted for her symptoms. From their sudden onset and the look of the MRI, Kenneth Smith, the hospital's chief of neurosurgery, suspected a fast-growing tumor called a glioblastoma. Glioblastomas, which afflict about 10,000 Americans each year, grow so rapidly — some doubling in size every ten days — that they cut off their own blood supply, leaving an area of dead tissue that appears dark against the illuminated mass on the MRI scan.

If it a glioblastoma, Perr's prognosis is grave. Surgery and radiation, the standard course of treatment, can give her a year or two at best. But this type of tumor inevitably returns. Glioblastomas literally crawl across the brain —

the cells can move — along the star-shaped glial cells that support and insulate networks of neurons. By the time a glioblastoma is diagnosed, it has usually spread so widely that even the most skillful surgeon cannot remove all of it.

However, Smith can offer Perr a ray of hope beyond surgery — an experimental treatment called GLI-328, which is now being tested at St. Louis University and 43 other hospitals in the United States and abroad. GLI-328 is neither radiation nor chemotherapy; it is a form of gene therapy, one that may offer victims of glioblastomas a chance to beat the odds against them.

The procedure will be performed in two parts by a team of three neurosurgeons. First, Smith and Lynn Bartel will remove as much of the tumor as possible. Then, if the diagnosis of glioblastoma is confirmed, Richard Bucholz will implant "suicide" genes into the remaining cancer cells to get them to self-destruct. Specifically, he will inject mouse skin cells carrying a genetically altered mouse leukemia virus into Perr's brain. The mouse virus contains a gene from the herpes simplex virus that causes cold sores, called the herpes thymidine kinase-1, or the herpes TK gene. Its manufacturer, Genetic Therapy, Inc., a Gaithersburg, Maryland, biotechnology company owned by Novartis Pharmaceuticals, has removed other bits of DNA to cripple the virus, so that it can infect but not reproduce in human cells. In particular, the virus can infect only dividing cells, which means that the only brain cells infected will be tumor cells, because normal brain cells do not divide. Once infected by the virus, the tumor cells will produce thymidine kinase, making them vulnerable to the antiviral drug ganciclovir.

"Surgery and radiation alone can't get all of the tumor cells. We need a smarter weapon," explains Bucholz, who developed a computerized-image guidance system that will play a crucial role in the surgery and the gene implantation. The surgical instruments used during the operation are equipped with infrared-light-emitting diodes, similar to those on a television remote control, and are continuously tracked by two overhead cameras. The information is then fed into a computer, which compares the position of the instruments with data collected and stored during the planning of the operation: a few days before, the team performed CT and MRI scans on Perr to visualize the tumor, distinguish the exact borders between malignant and normal tissue, and locate neighboring blood vessels. In the operating room a three-dimensional display based on this information will [provide the surgical team with a map to help plan the safest and most effective path to the tumor — and to deliver the gene therapy precisely to those areas where it should be needed most.

At 8:30 in the morning, the operation is under way. Bartel uses a small air-powered drill to remove a small rectangular section of Perr's skull. Then, with Smith holding open gelatinous flaps of brain tissue, she isolates and cauterizes blood vessels to cut off blood flow to the area of the tumor. Working quickly, guided by the computer images, the surgeons are able to remove the main portion of the tumor in less than an hour. A sample is rushed to the pathology lab for microscopic examination. Minutes later, the doctors' worst suspicions are confirmed: the tumor is indeed a glioblastoma.

In a lab on the floor below, medical technologist Tim Kilcoyne at once begins warming and preparing the deep-frozen virus-producing cells (VPCs) that will deliver the herpes TK gene to the tumor cells. The thawing process is delicate and slow. "The cells are easily damaged. You can't just pop them in the microwave," says Kilcoyne. His task is to count and test the viability of the cells over the next hour and a half. To produce enough herpes TK gene to combat the tumor requires 5 cubic centimeters (about a billion cells), enough to fill the bottom of a small plastic Ziploc bag.

While the cells are being prepared, Smith and Bartel try to remove a portion of the tumor that extends deeper into Perr's brain. By 11 o'clock, they have gone as far as they safely can.

In preparation for the gene therapy, Bucholz now places a meshwork grid made of suture threads and small metal clips across the tumor bed, inside the incision. His plan is to inject virus-producing cells at 1-centimeter intervals into each of the 20 squares of the grid, for maximum coverage. The procedure stalls when Kilcoyne calls to report that there are only 700 million cells in the preparation — fewer than the billion needed. The options are: use the available cells, begin the time-consuming work of preparing a new batch of cells, or inform Perr's family and give them the choice of declining the experimental treatment. Bucholz and Bartel cover the incision with a temporary dressing and wait.

A tense half hour later, officials at Genetic Therapy affirm by telephone that the cell count, though low, is still within the acceptable range. A few minutes later a nurse arrives with a cream-colored liquid in a plastic bag. Even after processing, the cells have a tendency to clump together, so she massages them gently, patting and smoothing the plastic packet.

Bucholz quickly loads the solution into a syringe with a long, springy needle. Then, using the computer images of Perr's brain to check position, direction, and depth, he begins the first infusion. Bartel presses the plunger of the syringe while Bucholz holds it in place. Rechecking the computer data, Bucholz then moves to the next position, repeating the process until the 5 cubic centimeters of solution have been distributed evenly over the grid. The entire procedure takes only ten minutes. Bucholz then removes the grid, and Bartel completes the five-hour operation by closing the incision, sealing Perr's skull with titanium strips, and stapling her skin closed.

After three days, Perr is discharged from the hospital. She will return in two weeks to have an intravenous line put into her arm for the twice-daily infusions of ganciclovir she will receive over the next two weeks. Ganciclovir is usually used to treat a disease related to herpes called cytomegalovirus, which causes blindness in patients with advance AIDS. Now she and the doctors hope the drug will slow down the tumor and possibly stop it from coming back.

Betty Perr is the seventh patient to receive gene therapy for a glioblastoma at the St. Louis University Health Sciences Center, and one of about 200 people around the world who have received GLI-328 in the five years since doctors started testing it.

The idea originated not as treatment but as a self-destruct mechanism for the first gene therapy experiment, explains Michael Blaese, chief of clinical gene therapy at the National Human Genome Research Institute, part of the National Institutes of Health. In 1990, Blaese was preparing to team up with gene therapy pioneer W. French Anderson in an attempt to transplant healthy genes into the blood cells of a four-year-old girl with severe combined immunodeficiency syndrome (SCIDS), the deadly "bubble boy" disease. The plan was to use a virus as a "vector," a microscopic truck to ferry the therapeutic genes into the girl's cells. "One of the viruses we were using was a mouse leukemia virus," recalls Blaese, "and there was a concern that if you inserted the genetic material into the wrong place, it could cause the cells to become cancerous. We wanted to build a self-destruct mechanism into these viruses. If the worst happened — if cancer occurred — we could pull the plug and kill the cells."

When laboratory experiments clearly demonstrated that infected cells producing the herpes TK gene could be killed by antiviral drugs, Blaese reflects, "it dawned on us that we could do it deliberately. Instead of using it as a fail-safe device, we could kill cancer with it."

More important, research in Blaese's lab showed that because the herpes TK gene changed ganciclovir into a chemical that was toxic to quickly dividing cells, the drug killed not only infected tumor cells but also nearby uninfected tumor cells, which were bathed in the toxin as the target cell died. Taking this "by-stander effect" into account, it seemed possible that as few as one in ten cancer cells would actually have to express the suicide gene for the treatment to work.

Using suicide genes as a weapon against cancer is far from the 1980s concept of gene therapy as a method to correct or replace the defective or missing genes responsible for deadly hereditary diseases such as SCIDS or cystic fibrosis. Yet applications like GLI-328 dominate gene therapy in the 1990s. According to the most recent records of the NIH's Recombinant DNA Advisory Committee, of 106 approved gene-therapy experiments, only a small fraction were aimed at correcting defective genes, while the vast majority involved inducing specific cells, such as cancer cells or cells infected by HIV, to produce proteins that would make them vulnerable to attack by the immune system or drugs.

From the beginning, glioblastomas seemed like ideal targets for such an approach. Glioblastomas kill by wreaking havoc on target areas in the brain, not by spreading throughout the body like other cancers. Therefore, a therapy would have to work only locally, not globally, to be effective. Another advantage is that normal brain cells do not divide. Therefore, since the mouse leukemia virus carrying the gene infects only dividing cells, only tumor cells will be targeted for destruction. Finally, the brain is an immunologically privileged site. To enter the brain, outside organisms must pass directly through the membranes of capillaries rather than through small clefts in their walls, as is the case throughout the rest of the body. Because the filtering effect of this blood-brain barrier keeps most dangerous agents away, the brain's immune defenses are weak and delayed. Elsewhere in the body, an infusion of virus-producing mouse cells would provoke the immune system to attack quickly, causing an inflammation in the area. In the brain, however, large numbers of mouse cells can be infused with only a small risk of causing inflammation.

In mice, GLI-328 completely killed tumors, without side effects. But while the results of the two human trials conducted so far are intriguing, they are also controversial. In 1992, Edward Oldfield and Zvi Ram, two neurosurgeons at the National Institute of Neurological Disorders and Stroke in Bethesda, Maryland, conducted a pilot study with their colleagues on ten patients who had undergone surgery for glioblastomas and whose tumors had come back. These patients received no repeat surgery to remove the regrowth. Instead the VPCs were delivered to the tumor site through a thin tube called a cannula, guided by CT-scan images. The patients then received follow-up treatments with intravenous ganciclovir.

In this small study GLI-328 was safe and effective in some cases, producing decreases in tumor size by 50 percent or more in four of the ten patients. The gene therapy helped, but not enough to produce any impact on survival. Because the delivery system limited the therapy to an area just a few cell layers deep (even with the bystander effect), only very small tumors responded. And while the therapy reduced parts of the tumors, distant portions progressed rapidly.

However, although nine of the ten glioblastoma patients died within three months to a year — about the same number expected from radiation or repeat surgery — one of the original patients, Kevin Klug, now 43, is defying the odds and is still alive five years later.

A new, improved clone of the herpes TK virus was introduced when a Phase II trial began at three hospitals in 1994. In this study 31 patients with recurrent glioblastomas underwent surgery to remove as much of the regrowth as possible, followed by a direct injection of GLI-328. A small plastic reservoir connected to a port on the skull's surface was implanted at the site of surgery to disperse subsequent injections if patients responded. (Some patients received as many as three treatments.)

Results of this trial have also been mixed, according to neurosurgeon, Mitchel Berger of the University of California at San Francisco, who was the lead investigator of the Phase II study. The average survival of the 31 patients was seven months — no improvement over repeat surgery and chemotherapy. But surprisingly, six of the patients — one out of five — are still alive three to four years later. "About one in a hundred people with glioblastomas gets better for some reason. But this is not some freak phenomenon. Twenty percent achieved a major benefit," Berger asserts. "We've never seen anything like it with any other therapy."

However, Edward Oldfield urges caution. "Despite the very small size of these tumors, we saw very little evidence of gene distribution — and even if we can reduce the size of the tumors by 50 percent, that's not enough to produce a lasting benefit," he points out. There could be other explanations for the longevity of the handful of long-term survivors, adds Oldfield, including sheer luck.

Is the gene therapy helping these patients beat impossible odds? The study now under way in St. Louis and elsewhere is trying to answer that question. During this Phase III trial, 250 patients in ten countries will receive either treatment with surgery, radiation, and gene therapy, or surgery and radiation alone. Unlike patients in earlier trials, those enrolled in the current study are receiving gene therapy at the time of their initial surgery, not at recurrence. They will be evaluated to see if the gene therapy reduces the time to recurrence, or increases survival compared with standard therapy. This is the first time any gene therapy has been evaluated in a large, randomized trial. If therapy with the herpes TK gene shows a benefit, it could be approved for widespread use as early as the year 2000.

The stakes are high. If successful, suicide genes could be applied to other forms of cancer, or to stop restenosis, the rapid cell growth responsible for reclosing heart vessels after balloon angioplasty. Researchers are also considering using suicide gene strategies to control graft-versus-host disease after bone marrow transplantation and to kill off the proliferating cells that cause inflamed arthritic joints.

Elsewhere, doctors are testing similar techniques to fight disease. Recently, Jeffrey Isner of St. Elizabeth's Medical Center in Boston reported dramatic success in ten patients who had such severe peripheral artery disease that they were in danger of having their legs amputated. After Isner implanted genes for a hormone that stimulates the growth of blood vessels, most of the patients sprouted collateral arteries that bypassed their blocked arteries. Nine out of ten patients improved so markedly that amputations were avoided or limited to a toe.

Results with these therapies will have to overcome serious doubts about whether gene therapy will fulfill its seemingly vast potential. According to a widely discussed review of the field commissioned two years ago by NIH director Harold Varmus, after more than 100 clinical gene therapy trials at a cost of well over $200 million a year, there was still no unambiguous evidence that genetic treatment has produced therapeutic benefits.

Will GLI-328 be an exception, the first gene therapy for cancer to demonstrate clear-cut results? For Richard Bucholz the treatment represents an exciting step forward, whether or not it produces dramatic benefits against brain tumors. "We're just starting on the road to establishing the techniques to get the maximum response out of gene therapy. If we are successful, this approach represents a major shift in the way we treat disease," he reflects.

"The gene therapy approach is very exciting because it bypasses the natural biology of these tumor cells, which are very difficult to understand and treat," adds UCSF neuro-oncologist Michael Prados, who tested GLI-328 in the Phase II trial. "It allows you to create a tumor cell that's the way you want it to be — killable." ❑

Editor's note: *However promising GLI-328 may be for other glioblastoma patients, it was unable to save Betty Perr from her cancer. Sadly, just as we were about to go to press, we learned that this courageous woman had died. We extend to her family our condolences and our gratitude for their generous cooperation in preparing this article.*

* * * * * * * * * * * * * * * * * * * *

Questions:

1. What role do "suicide genes" have in gene therapy? *Cause apoptosis in target cells*

2. How do viruses help scientists perform gene therapy procedures? *Serve as vectors*

3. Describe the two main categories of all the NIH approved gene-therapy experiments to date.

4. Why is gene therapy believed to be more effective in treating glioblastomas than other types of cancer?

Answers are at the end of the book.

Investigators in Canada studied over 1,896 Caucasian Canadian patients with multiple sclerosis (MS) and almost 9,000 of their first-degree relatives in hopes of determining the susceptibility of MS within families. The group found that MS susceptibility within families may be more accurately determined from the age of disease onset and sex of affected siblings and the paternal disease status. The group found that brothers of male patients have a higher rate of MS than brothers of female patients, whereas sisters of male patients had a similar rate of MS to sisters of female patients. An earlier age at onset also affected lifetime risk, so that siblings of a patient who had developed MS at age 20 or younger were almost 5 times more likely to develop MS than siblings of patients whose age at onset was older than 40 years. This study's findings are significant because they provide specific age-adjusted figures for MS susceptibility, which will allow more precise genetic counseling.

Effect of Age at Onset and Parental Disease Status on Sibling Risks for MS

by A. Dessa Sadovnick, *et al.*

Neurology, March 1998

Article abstract — *MS is believed to be a complex trait determined by genetic and nongenetic factors. Data suggest that MS susceptibility and age at onset are each, at least to some extend, under genetic control. The present study carefully examined five covariates (sex of the index case, sex of the sibling, birth cohort of the sibling [≤1919, 1920 to 1939, ≥1940], age of MS onset in the index patient (≤20 years, 21 to 30 years, 31 to 40 years, >40 years), and MS disease status of the parents [i.e., MS present in one parent or no parent with MS]) that may influence the familial risk of MS in a large cohort of 1,896 MS patients and 8,878 of their first-degree relatives. Of these, sex of the sibling, parental MS status, and index patient onset age were the important factors influencing MS risks to siblings. The results of this study are (1) the index-patient-onset-age effect suggests that individuals with a greater genetic loading (i.e., a greater contribution of susceptibility alleles) have an earlier age at onset and (2) genetic loading is substantially increased in individuals with an affected parent. These data are important both for genetic counseling and gene identification studies.*

Several lines of evidence support the notion that MS is a complex trait determined by genetic and nongenetic factors. This evidence comes from studies of twins,[1-3] "adoptees,"[4] half-siblings,[5] and familial recurrence risk data.[4,6,7] The familial aggregation of MS appears to be caused by genetic factors.[4,5]

The distribution of onset ages for MS varies widely, spanning more than five decades.[8] Studies have shown that the ages at onset for concordant monozygotic twins are much more strongly correlated that the ages at onset for concordant nontwin siblings, which in turn are more correlated than ages of onset for unrelated, age-matched MS patients.[9] (Concordant dizygotic pairs are too rare to allow such meaningful analysis.) Taken together, these findings suggest that MS susceptibility and age at onset are each, at least to some extend, under genetic control.

The steep drop in recurrence risks between monozygotic twins and dizygotic twins or nontwin siblings, and also for more distant relatives, strongly suggests an oligogenic or polygenic disease model.[2-4, 6,7] This is also a plausible explanation for the observation that recurrence risks for first-degree relatives of concordant monozygotic twins exceed those for first-degree relatives of discordant monozygotic twin pairs.[1,10]

The present study carefully examines factors that may influence the familial risk of MS in a large cohort of MS patients and their first-degree relatives.

Methods. *MS index patients.* Since September 1980, all consecutive, unrelated patients attending the Vancouver MS Clinic (hereafter referred to as the MS Clinic) have had detailed family histories taken and documented. Family information is collected from the index patients and other family informants as appropriate. Family histories are updated on an annual basis. This database represents as unbiased a population as possible with respect to familial MS.[6] Care is taken to identify and exclude index patients who attend the MS Clinic specifically because of concerns about familial MS from the "population-based" database.

The present study includes all Caucasian MS Clinic patients (index patients) diagnosed with clinically probable/clinically definite and/or laboratory-supported definite MS seen at the MS Clinic from September 1980 to January 1986.[8,11] "Adopted" and non-Caucasian MS index patients were excluded from this study, as were offspring of conjugal MS parents.

Diagnostic criteria for MS first-degree relatives of index patients. Whenever possible, parents and siblings of index patients reported to be affected by history were examined at the Vancouver (or other Canadian) MS Clinic. If this was impossible (usually because of death or geographic location outside of Canada), appropriate physician and/or hospital (autopsy) records were obtained and reviewed by MS Clinic personnel with expertise in the differential diagnosis of MS. Parents and siblings of index patients were considered to be affected only if they met the criteria for definite, clinically probable/clinically definite, or laboratory-supported definite MS.

An individual was excluded from the analyses if definitive information could not be obtained on age at present (or age at death) and/or disease status (MS present in one parent or no MS in either parent).

Data analysis. Average age at onset by seven age groupings (≤20, 21 to 30, 31 to 40, 41 to 50, 51 to 60, 61 to 70, and >70) were computed for all index patients and their relatives for whom present age (or age at death) and age at MS onset were known. For affected relatives with missing age-at-onset information but known present ages (or ages at death), the average age at onset using the seven age groupings was used. For each subgroup defined by a covariate of interest, risks to siblings of MS index patients were first calculated using Kaplan-Meier life table methods.[12] To investigate the joint relationship of these covariates to the sibling MS risks, data were subsequently analyzed using Cox proportional hazards analysis.[13]

The Cox proportional hazards analysis assumes that (1) a baseline hazard function exists; (2) different individuals have hazard functions that are proportional to one another, determined by covariates; and (3) covariates have a multiplicative effect on the hazard function. The hazard function is the instantaneous rate of disease occurrence at time t, given that the individual is disease free until time t.

The following five covariates were considered in the Kaplan-Meier and Cox proportional hazards analyses: (1) sex of the index patient, (2) sex of the sibling, (3) birth cohort of the siblings (≤1919, 1920 to 1939, ≥1940), (4) age at MS onset in the index patient (≤20 years, 21 to 30 years, 31 to 40 years, >40 years), and (5) MS disease status of parent (i.e., MS present in one parent, no parent with MS).

Results. Of the 2,156 MS index patients, 260 (12.1%) were excluded from this analysis because of either incomplete information on parents and/or siblings or the index patients was an only child. Of the 1,896 index patients included, nine (three males and six females) did not have age-at-onset data available. For these nine patients, age at onset was estimated by the average age at onset of the other 1,887 index patients in the seven age groupings listed previously.

Complete data on age at present (or age at death) and/or disease status (MS present, no MS) were available for 8,878 parents and siblings of the 1,896 Caucasian MS index patients. Complete information was available for 97.3% of fathers (1,844 of 1,896) and 98.8% of mothers (1,874 of 1,896). For most of the parents, the missing information was usually the exact age of death. Ninety of the 5,160 siblings (23 brothers and 67 sisters) were affected with MS. Age at onset was not known for 27 of these 90 affected siblings (30.0%; seven brothers and 20 sisters). For these 27 siblings, it was necessary to assign average age at onset arbitrarily (based on the seven age groupings) based on known ages at onset and ages at present (or at death) for the other 70.0% of affected siblings (n = 63). There is inherently more variability in this age-of-onset distribution than we have assumed. Nevertheless, given the relatively small proportion of siblings with missing age-at-onset information, it is unlikely that this adjustment altered significantly the patterns we observed.

The following summarizes the distribution of siblings by sex and age at MS onset in the index patient. In general, females tend to have a younger age at MS onset than males. Thus, a greater proportion of siblings of female index patients falls into the ≤ 30-index-patient-onset-age group than do siblings of male index patients. However, these differences are not dramatic and do not confound the analyses. Overall, the age-unadjusted MS rate is higher in brothers of male index patients than brothers of female index patients (1.5% vs. 0.6% $X^2 = 3.98$, $p < 0.05$). The opposite trend is observed for sisters, although differences are not significant statistically. The age-unadjusted MS

rate tends to be similar in sisters of female index patients compared with sisters of male index patients (2.7% versus 2.3%).

The index patient onset age effect is more apparent for sisters than for brothers. For sisters, the age-unadjusted MS rates are 4.5%, 1.7%, 2.3%, and 1.4% for the four index patient onset age groups in ascending order (≤ 20, 21 to 30, 31 to 40, >40 years). For brothers, the comparable rates are 0.9%, 0.9%, 1.2%, and 0.6%. However, the siblings of index patients with a younger age at onset tend to be younger themselves, and so these rates need to be age adjusted (as discussed later).

The following shows the distribution of siblings by sex of index patient and parental MS status. Combining data for male and female index patients, the risk to siblings with one affected parent is substantially greater than when there is no affected parent. For brothers having no affected parent the age-unadjusted risk is 0.8%, whereas for brothers with one affected parent the risk is 3.4% (2 of 59). For sisters with no affected parent the risk is 2.4% whereas for sisters with one affected parent the risk is 8.1% (6 of 74). The increased risk associated with having an affected parent is more apparent for fathers than for mothers. For brothers with an affected father the risk 8.7% (2 of 23), whereas for brothers with an affected mother the risk is 0% (0 of 36). Similarly, for sisters with an affected father, the risk is 16.7% (3 of 18) compared with 5.4% (3 of 56) for sisters with an affected mother. Due to small numbers, these comparisons are not formally significant, but are highly suggestive. Particularly intriguing are the two affected brothers of male index patients with affected fathers.

This observation is somewhat unexpected given the lower MS rate in males than in females, but is in agreement with our previous observation that father-son pairs concordant for MS tend to come from more "loaded" MS families.[14]

Age-adjusted lifetime risks for MS were calculated by the Kaplan-Meier life table method for subgroups of subjects defined by the covariates. Risk curves suggested that sex of the sibling, parental MS status, and index patient onset age were the important factors influencing MS risks to siblings. Curves confirm the age-unadjusted analyses. The lifetime risk for sisters of index patients is about 2.5 times that for brothers. The lifetime risk for siblings with an affected parent is about fourfold greater than for those without an affected parent. The lifetime risk is greatest for siblings of index patients with an MS age at onset ≤20 years and it decreases continuously, about fourfold, as the index patient onset age increases. It is also apparent that nearly all the early-onset patients (younger than 30 years) among the siblings occur in families of index patients with early onset (younger than 30 years), providing further evidence[9] of a correlation in index patient and sibling age of onset.

The joint relationship of the five covariates to the sibling risks was examined using the Cox proportional hazards analysis. Sex of the index patient and birth cohort of the sibling were not significant and thus were subsequently dropped as covariates for the analyses. The significant covariates were sex of the sibling, parental MS status, and index patient onset age. The different combinations of interaction terms of these three covariates were not significant. There are hazard ratios for the three significant covariates.

There are MS lifetime risk percentages for siblings of MS index patients based on the Kaplan-Meier estimates for each of these categories. MS risks for both brothers and sisters increase with earlier index patient onset age and one parent with MS. The MS risks are consistently increased for sisters compared with brothers. There are also fitted lifetime risk percentages for siblings obtained from proportional hazards analysis using various parameters. The fitted values give reasonably close correspondence to the observed values. Some of the standard errors for the observed values are large due to small sample sizes, particularly in the case of brothers with one affected parent.

Discussion. Genetic epidemiologic data strongly indicate that MS is a complex trait determined by multiple genes and nongenetic (environmental) factors. The prevalence of MS has been shown to vary by gender, age, geography, and ethnic background. The ethnic risk, regardless of place,

appears to be accounted for by susceptibility genes, but the number of loci, mode of inheritance, and mechanism of inheritance remain largely unknown.

Previously published familial risk data[4-7] could not divide families according to covariates because of the relatively small sample sizes. Previous MS Clinic data give the overall lifetime risk for siblings of Caucasian MS index patients also to develop MS (data for 2,138 siblings of 815 index patients) as $3.6 \pm 0.50\%$.[1,6] The larger number of index patients (more than a 2.3-fold increase over the previous data set of 815 index patients) in the present study allows more precise estimates of risk, which take into account the important covariates of sex of the sibling. These data are relevant to genetic (risk) counseling, because information about index patient onset age and parental MS status can usually be obtained from the consultand.

The risk differential for siblings of MS index patients stratified by index patient onset age and parental MS status is dramatic. According to the proportional hazards analysis, having one affected parent increases the sibling hazard 3.5-fold. The increased risk associated with affected fathers may be even greater, but small sample size did not allow us to support this observation strongly. The hazard analysis also reveals that the risk to the siblings of the earliest onset patients (≤20 years) may be up to fivefold higher than for siblings of the oldest onset patients (>40 years).

These observations have two important implications regarding the genetic basis for MS. First, the index-patient-onset-age effect suggests that individuals with a greater genetic loading (i.e., a greater contribution of susceptibility alleles) have an earlier age of onset. Second, the strong effect of having one affected parent suggests genetic variability in risk whereby genetic loading is substantially increased in individuals with an affected parent.

Neither observation, however, can delineate clearly between major gene and multigenic models. It is well understood that for multigenic models, risks to relatives increase when the index patient has a positive family history or an early age at disease onset. On the other hand, models allowing for a single major gene as well as sporadic cases can also produce similar effects. For example, the genetics of breast cancer is characterized primarily by two loci[15] (BRCA1 and BRCA2) with rare mutations that induce higher risk at early ages, but account for only a small proportion of all breast cancer. Sisters of breast cancer index patients with an earlier age of onset and/or and affected mother are at a substantially increased risk compared with other women.

In summary, the observations reported in this paper for MS are important both for genetic counseling and gene identification studies. More precise familial recurrence estimates can be given based on age at onset and family history of the index patient. This is most likely to occur in the setting of an affected consultand with a family history who request risks for siblings and offspring. While such risk estimates have not yet been tabulated for offspring because of the average youth of this group, results suggest that these risks would likely be greater if the index patient had an early age at onset and/or a positive family history.

Three genome screens using linkage analysis to analyze MS susceptibility genes have been published.[16-18] These studies revealed no consistent findings for any particular chromosomal region to be linked to MS susceptibility, except for the HLA region on chromosome 6p. The data presented here suggest that it may be useful to stratify families used in linkage studies by number of affecteds and age at onset, because some subsets may be more powerful in revealing linkage evidence than others.

The results of these analyses further emphasize the role of genetic factors in the familial aggregation of MS. They are in agreement with the previous suggestion by our group[9] that age at MS onset may be, at least to some extent, under genetic control. Finally, the combined effect of age at MS onset, gender, and parental MS status on sibling risks are what would be predicted under a liability threshold model for MS susceptibility. The data suggest that the age at MS onset differentiates earlier onset MS patients by the relative load of susceptibility factors, which seem likely to be genetic factors.

The data presented in this paper can be used for more precise genetic counseling purposes as well as for genetic modeling of MS susceptibility. ❑

References

1. Ebers GC, Bulman DE, Sadovnick AD, et al. A population-based twin study in multiple sclerosis. *N Engl J Med* 1986; 315:1638-1642.

2. Sadovnick AD, Armstrong H, Rice GAP, et al. A population-based twin study of multiple sclerosis in twins: update. *Ann Neurol* 1993; 33:281-285.

3. Mumford CJ, Wood NW, Kellar-Wood H, Thorpe JW, Miller DH, Compston DA. The British Isles survey of multiple sclerosis in twins. *Neurology* 1994; 44:11-15.

4. Ebers GC, Sadovnick AD, Risch NJ, Canadian Collaborative Study Group. Familial aggregation in multiple sclerosis is genetic. *Nature* 1995; 377:150-151.

5. Sadovnick AD, Ebers GC, Dyment D, Risch NJ, Canadian Collaborative Study Group. Evidence for the genetic basis of multiple sclerosis, *Lancet* 1996; 347:1728-1730.

6. Sadovnick AD, Baird PA, Ward RH. Multiple sclerosis: updated risks for relatives. *Am J Med Genet* 1988; 29:533-541.

7 Robertson NP, Fraser M, Deans J, Clayton D, Walker N, Compston DAS. Age-adjusted recurrence risks for relatives of patients with multiple sclerosis. *Brain* 1996; 119:449-455.

8. Poser CM, Paty DW, Scheinberg L, et al. New diagnostic criteria for multiple sclerosis: guidelines for research protocols. *Ann Neurol* 1983; 13:227-231.

9. Bulman DE, Sadovnick AD, Ebers GC. Age of onset in siblings concordant for multiple sclerosis. *Brain* 1991; 114:937-950.

10. Mackay RP, Myrianthopoulos NC. Multiple sclerosis in twins and their relatives. Final report. *Arch Neurol Psychol* 1966; 15:449-462.

11. Schumacher GA, Beebe G, Kibler RF, et al. Problems of experimental trials of therapy in multiple sclerosis. Report by the panel on the evaluation of experimental trials of therapy in multiple sclerosis. *Ann NY Acad Sci* 1965; 122:552-568.

12. Kaplan EL, Meier P. Non-parametric estimations from incomplete observations. *J Am Stat Assoc* 1958; 53:457-481.

13. Lawless JF. Statistical models and methods for lifetime data. New York: Wiley, 1982: 343-400.

14. Sadovnick AD, Bulman D, Ebers GC. Parent child concordance in multiple sclerosis. *Ann Neurol* 1991; 29:252-255.

15. Claus EB, Schildkraut JM, Thompson WD, Risch NJ. The genetic attributable risk of breast and ovarian cancer. *Cancer* 1996; 77:2318-2324.

16. Ebers GC, Kukay K, Bulman DE, et al. A full genome

screen in multiple sclerosis. *Nat Genet* 1996; 13:472-476.

17. Sawcer S, Jones HB, Feakes R, *et al*. A genome screen in multiple sclerosis reveals susceptibility loci on chromosome 6p21 and 17q22. *Nat Genet* 1996; 13:464-468.

18. Multiple Sclerosis Genetics Group. Complete genomic screen for multiple sclerosis underscores a role for the major histo-compatibility complex. *Nat Genet* 1996; 13:469-471.

* * * * * * * * * * * * * * * * * * * *

Questions:

1. What were the five covariates examined in this study? *1) Sex of index case 2) sex of siblings 3) birth cohort 4) age of onset of MS case 5) MS disease status of parents*

2. True or False? Multiple sclerosis is seen more frequently in females.

3. What were the three greatest factors that influenced MS risks to siblings? *1) Sex 2) parental MS status 3) index patient onset age*

4. To date, what is the only chromosomal region to be linked to MS susceptibility? *HLA, 6p*

Answers are at the end of the book.

- 21 -

Some people thrive on exercise. Others, unfortunately, the majority of the population, are couch potatoes, ex-members of one or more gyms, discouraged by their inability to get as buff as they want or lose those few extra pounds. Wouldn't it be great if there was a way to tell before handing over those membership fees whether all that sweat will do any good? Researchers in London have found a genetic factor associated with physical endurance and positive responses to physical training regimes. The research focuses on an enzyme called angiotensin-converting enzyme (ACE) which comes in two forms, "D" or "I." The group looked at 123 consecutive white male army recruits to see how their stamina improved after a 10-week training regimen. Blood samples were taken before and after the training period and compared to their ACE genotype. Preliminary results suggest that those of us who are homozygous for the D allele may have a legitimate excuse for not putting ourselves out in the gym.

Human Gene for Physical Performance

by H.E. Montgomery, *et al.*

Nature, May 12, 1998

A specific genetic factor that strongly influences human physical performance has not so far been reported, but here we show that a polymorphism in the gene encoding angiotensin-converting enzyme does just that. An 'insertion' allele of the gene is associated with elite endurance performance among high-altitude mountaineers. Also, after physical training, repetitive weight-lifting is improved eleven-fold in individuals homozygous for the 'insertion' allele compared with those homozygous for the 'deletion' allele.

The endocrine renin-angiotensin system is important in controlling the circulatory system. Angiotensin-converting enzyme (ACE, or kininase II) degrades vasodilator kinins, and converts angiotensin I (ATI) to the vasoconstrictor angiotensin II (ATII). In addition, local renin-angiotensin systems may influence tissue growth[1]. A polymorphism of the human *ACE* gene has been described in which the deletion (*D*) rather than insertion (*I*) allele is associated with higher activity by tissue ACE[2].

There is evidence for a skeletal muscle renin-angiotensin system[3], suggesting that muscle growth, and thus physical performance, might be positively associated with the *D* allele. However, our initial studies suggested that the *I* allele was associated with improved endurance performance. We investigated this association in two parallel experiments.

High-altitude mountaineers perform extreme-endurance exercise. Thirty-three elite unrelated male British mountaineers, with a history of ascending beyond 7,000 metres without using supplementary oxygen, were identified by the British Mountaineering Council. DNA was extracted from a mouthwash sample of the 25 male respondents, and *ACE* genotype was determined using a three-primer polymerase chain reaction amplification[4].

Genotype distribution was compared with that of 1,906 British males free from clinical cardiovascular disease[5]. Mean age was 40.6 ± 6.5 years in the 25 subjects, and 55.6 ± 3.2 years for the controls. Both groups were in Hardy-Weinberg equilibrium. Both genotype distribution and allele frequency differed significantly between climbers and controls (P was 0.02 and 0.003 respectively (X^2 test)), with a relative excess of II genotype and deficiency of DD genotype found in the climbers.

Among the 15 climbers who had ascended beyond 8,000 m without oxygen, none was homozygous for D (6 II and 9 ID: I allele frequency - 0.65). Further, ranked by number of ascents above 8,000 m without oxygen, the top performer was homozygous for I (5 ascents, compared with a mean of 2.4 ± 0.3 ascents, or 1.44 ± 0.3), as were the top two in this group for number of additional 7,000-m ascents (> 100 and 18, compared with a mean of 10.3 ± 6.5 ascents).

In a second study, ACE genotype was determined in 123 Caucasian males recruited to the UK army consecutively. Seventy-eight completed an identical 10-week general physical training programme (age, 19.0 ± 0.2 years; height, 176.6 ± 0.7 cm; body mass index, 22.2 ± 0.2 kg m^2). Their ACE genotype (20 (25.6%) II, 46 (59.0%) ID, 12 (15.4%) DD) matched that of those who failed training, as did their physical characteristics (neck, chest and waist circumference, elbow diameter and armspan), and all characteristics were independent of genotype.

The maximum duration (in seconds) for which they could perform repetitive elbow flexion while holding a 15-kg barbell was assessed both before and after the training period. Pre-training performance was independent of genotype (mean, $119.8 + 6.2$ s). Duration of exercise improved significantly for those (66 individuals) of II and ID genotype (79.4 ± 25.2 and 24.7 ± 8.8 s: P was 0.005 and 0.007 respectively) but not for the 12 of DD genotype (7.1 ± 14.9 s: $P = 0.642$). Improvement was thus eleven-fold greater ($P = 0.001$) for those of II than for those of DD genotype.

Genotype-dependent improvements were unlikely to be due to changes in individual muscle fibre size and strength (which need more than three months of specific strength-training to occur) or

altered co-ordination, neural firing pattern or recruitment of fast motor units (given the lack of specific training for the test task)[6-8]. Increased performance is therefore most likely to be due to an improvement in the endurance characteristics of the tested muscles.

The association of the I allele with improved endurance might derive from variable increases in substrate delivery due to increases in cardiac output and muscle capillary density; from changes in the nature of substrate used, due to a differential shift to stored fatty acids as fuel[9], or in the efficiency of substrate utilization relating to altered muscle fibre type; from altered mitochondrial density, or from raised muscle myoglobin content [10, 11]. Elevated catecholamine, cortisol and growth hormone concentrations may all increase the availability of oxidative fuel[12].

Further work will be needed to determine whether this correlation holds beyond the limited group studied here and to explore the mechanisms underlying these observations. ❑

References
1. Katz, AM. *Heart Disease and Stroke* 1, 151-154 (1992).
2. Danser, AH, *et al. Circulation* 92, 1387-1388 (1995).
3. Reneland, R. & Lithell, H. *Scand. J. Clin. Lab. Invest.* 54, 105-111 (1994).
4. Montgomery, HE, *et al. Circulation* 96, 741-747 (1997).
5. Miller, GJ, Bauer, KA, Barzegar, S, Cooper, JA & Rosenberg, RD. *Thromb. Haemost.* 75, 767-771 (1996).
6. Komi, PV. in *Biochemistry of Exercise VI* (ed. Saltin, B) 529-574 (Human Kinetics, Champaign, Illinois, 1986).
7. Rutherford, OM, Greig, CA, Sargeant, AJ, & Jones, DA. *J. Sports Sci.* 4, 101-107 (1986).
8. Jones, DA & Rutherford, OM. *J. Physiol.* 391, 1-11 (1987).
9. Rennie, MJ, Winder, WW, & Holloszy, O. *Biochemistry* 156, 647-655 (1976).
10. Bloom, SR, Johnson, RH, Park, DM, Rennie, MJ & Salaiman, WR. *J. Physiol.* 258, 1-18 (1976).
11. Hudlicka, O. *News Physiol. Sci.* 3, 117-120 (1988).
12. Wasserman, DH & Vranic, M. in *Biochemistry of Exercise VI* (ed. Saltin, B.) 167-179 (Human Kinetics, Champaign, Illinois, 1986).
13. SAS Institute. *SAS Users Guide* (SAS Institute, Cary, North Carolina, 1985).

* * * * * * * * * * * * * * * * * * * *

Questions:

1. True or False? According to this study, the *I* allele is associated with a positive effect on physical performance.

2. How did the duration of exercise differ between the males identified as having an *II*, *ID*, and *DD* genotype?

3. What might contribute to the genotype-dependent improvements seen in this study?

Answers are at the end of the book.

Part Six

Mutation,
Recombination,
and
Repair

Caenorhabditis elegans is a nematode — members of the phylum Nematoda — that has been studied extensively in genetics and evolutionary biology. The Nematoda phylum is made up of roundworms and threadworms, a phylum of smooth-skinned, unsegmented worms with a long cylindrical body shape tapered at the ends; includes free-living and parasitic forms both aquatic and terrestrial. C. elegans is about as primitive an organism that exists which nonetheless shares many of the essential biological characteristics that are central problems of human biology. The worm is conceived as a single cell that undergoes a complex process of development, starting with embryonic cleavage, proceeding through morphogenesis and growth to the adult. C. elegans is comprised of two sexes, however, one sex, the male, encoumbers the sexually reproducing species more than the production of females. While most scientists have been studying their ideas for balancing the twofold cost of producing males in this organism, a group of researchers having been studying the cousin of C. elegans, the Caenorhabditis briggsae. These evolutionary biologist have discovered that C. briggsae has found a unique way of gaining all the benefits of sexual reproduction without paying the cost of producing males.

How to Gain the Benefits of Sexual Reproduction without Paying the Cost: A Worm Shows the Way

by Raghavendra Gadagkar, *et. al.*

Trends in Ecology & Evolution, June 6, 1998

Sexual reproduction is perhaps the greatest of all evolutionary puzzles. It's a puzzle because sexually reproducing species pay the cost of spending half their resources (over and above what is needed for vegetative growth) in producing males, whereas parthenogenetic species utilize all their resources meant for reproduction in producing only females (or hermaphrodites) like themselves. This twofold cost of sexual reproduction[1,2] is sometimes referred to as the twofold cost of producing males. Three advantages of sexual reproduction that might offset this cost have been proposed. Genetic recombination and cross fertilization permit sexually reproducing species to (1) bring together, in the same individual, mutations arising in different individuals[3,4]; (2) generate genetic variability and thus adapt to changing environments[2,5,6]; and (3)

shuffle their genes in every generation and thus keep parasites at bay[7-9]. While evolutionary biologists are busy testing their favourite ideas for offsetting the twofold cost of producing males, recent work by Craig LaMunyon and Samuel Ward[10] shows that a nematode, *Caenorhabditis briggsae*, appears to have found a way of gaining the benefits of sexual reproduction without paying the cost of producing males.

Before we understand how the worm accomplishes this, it would be useful to familiarize ourselves with its more famous cousin, *C. elegans* — a remarkable nematode that was chosen by geneticists after a deliberate search for a simple multicellular eukaryote[11,12]. It is 1 mm long and has <1000 cells in its body. Most individuals are hermaphrodites, which produce both oocytes and sperm and reproduce by self-fertilization. Males are occasionally produced because of nondisjunction of the X chromosomes. Hermaphrodites have two X chromosomes and hence their oocytes and sperm each carry a single X chromosome. Males carry a single copy of the X chromosome and are designated XO. Their sperm may either carry a single copy of X or may carry no copies — the latter are designated nullo-X. Occasionally, hermaphrodites mate with males and produce outcrossed male and hermaphrodite offspring. Perhaps because mating is so rare, the worm makes the best use of the opportunity available and produces nearly all outcrossed progeny. Both the X-bearing and nullo-X sperm of the male take precedence over the sperm of the hermaphrodite, so that roughly equal proportions of outcrossed male and outcrossed hermaphrodite offspring are produced. Thus, even *C. elegans* pays the twofold cost of producing males to gain the advantage of sexual reproduction. However, given the advantage of outcrossing, it should be expected that both the male and the hermaphrodite cooperate to suppress the hermaphrodite sperm — the male can preferentially transmit his genes and the hermaphrodite gains the advantage of sexual reproduction.

LeMunyon and Ward[10] have compared *C. elegans* with *C. briggsae* and found that the latter adopts an even better strategy. During the early period after mating, *C. briggsae* hermaphrodites produce only ~10% male offspring instead of the expected 50%. The remaining 90% hermaphrodite offspring are nearly all outcrossed progeny. This was revealed by a trick employed by the experimenters — although the males were wild type, the hermaphrodites used for mating were homozygous for a recessive mutation that produces short and fat worms, appropriately referred to as 'chubby'. Thus, hermaphrodite progeny resulting from self-fertilization would be 'chubby', but those resulting from outcrossing would be wild type. The observation that only about 10% of the offspring were male and the remaining 90% of offspring were all wild-type hermaphrodites suggests that the X-chromosome-bearing sperm of the males took precedence not only over the sperm of the hermaphrodite (as in *C. elegans*), but also over the

nullo-X sperm of the males (unlike *C. elegans*). Therefore, when presented with an opportunity to mate and outcross, *C. briggsae* worms gain the benefits of sexual reproduction without paying the cost of producing males.

It is clear that this process concerns competition between sperm rather than the enhanced survival of outcrossed hermaphrodite offspring relative to ourcrossed male offspring (mortality of progeny was found to be independent of their gender). It is also clear that sperm competition is not a result of numerical dominance — another unusual, but very useful character of *Caenorhabditis* is their extreme economy of sperm production and use. The numbers are so small and their use is so efficient that LaMunyon and Ward were able to account for all the sperm and show that male sperm can outcompete hermaphrodite sperm even when they are outnumbered by two to one. Even during selfing, each sperm is used to fertilize one oocyte. In contrast to most other organisms, oocytes are made in excess and some of them can remain unfertilized. As they mature, the oocytes reach the entrance of the spermatheca, where the lumen opens to release sperm. The oocyte then appears to 'ingest' a single sperm (fertilization) and excess sperm get swept into the uterus but subsequently crawl back into the spermatheca to await future ingestion[13].

LaMunyon and Ward have ruled out an alternative explanation for their results — that delayed fertilization is an inherent property of all nullo-X sperm. If this was the case, the interpretation that delayed fertilization represents an adaptation to gain the benefits of sexual reproduction without paying the cost loses much of its strength. Using a mutant *C. briggsae* that produces some nullo-X bearing sperm (owing to increased nondisjunction), they showed that the nullo-X sperm produced by the hermaphrodite are not inferior to their X-bearing counterparts.

The male X-bearing sperm exclude not only the hermaphrodite sperm, but also the nullo-X sperm of the male. The X chromosome of the male can, therefore, be described as a selfish chromosome because it enhances its own survival at the expense of other chromosomes (present in the nullo-X sperm) in the same individual. However, again, it should be expected that hermaphrodites cooperate in this conspiracy because they can then gain cost-free benefits of sexual reproduction. The selfishness of the X chromosome of the male is only expressed when it passes through a male and not when it passes through a hermaphrodite. Such a phenomenon, in which the same chromosome behaves differently depending on whether it has a paternal or maternal origin, is known as imprinting[14-16].

The selfish X chromosome of *C. briggsae* is reminiscent of the selfish B chromosome of the parasitic wasp *Nasonia vitripennis*[17,18] (which oviposits in the pupae of certain flies). Being a hymenopteran, it follows a haplodiploid life cycle — males develop from unfertilized eggs and are haploid, and females develop from fertilized eggs and are diploid. In some strains, males harbour a B chromosome that has been described as the most selfish genetic element known. The B chromosome confers no known advantage to the male who carries it. Instead, it reaches the zygote via the sperm and inactivates all the paternal chromosomes that have come with it. The diploid zygote, which might have otherwise developed into a female, is thus converted into a male haploid cell consisting only of maternal chromosome (plus the B chromosome) thus ensuring the continued propagation of the B chromosome. The selfish B chromosome of *N. vitripenis* differs from the selfish X chromosome of *C. briggsae* in several ways: the selfish B chromosome destroys its competitor chromosomes, whereas the selfish X chromosome only delays their fertilization; the B chromosome is always selfish, but the selfishness of the X chromosome depends on imprinting; the selfish B chromosome favours the production of sons, whereas the selfish X chromosome delays or prevents the production of sons; and the selfish B chromosome confers no advantage either to the male that carries it or to the female whose oocyte it enters, whereas the selfish X chromosome confers a clear advantage to the female whose oocyte it enters.

In spite of these differences between the selfish X chromosome of *C. briggsae* and the selfish B

chromosome of *N. vitripennis*, their behaviour demonstrates that the broad rules of cooperation and conflict are as applicable to a set of interacting biological molecules as they are known to be to a set of interacting organisms or populations[18,19]. However, it is remarkable that the strategy of *C. briggsae*, which gains all the advantages of sexual reproduction without paying any of the cost of producing males, is not more widespread. Perhaps it comes with a cost that we do not yet appreciate or perhaps it is more widely utilized than we are aware — after all, only a tiny proportion of the described species, let alone the undescribed ones, have been studies to an appreciable degree. ❏

References

1. Maynard Smith, J. (1971) **The origin and maintenance of sex**, in *Group Selection* (Williams, G.C., ed.), pp. 163-175, University of Chicago Press
2. Maynard Smith, J. (1978) *The Evolution of Sex*, Cambridge University Press
3. Fisher, R.A. (1930) *The Genetical Theory of Natural Selection*, Oxford University Press
4. Muller, H.J. (1932) **Some genetic aspects of sex**, *Am. Nat.* 66, 118-138
5. Williams, G.C. (1975) *Sex and Evolution*, Princeton University Press
6. Bell, G. (1982) *The Masterpiece of Nature: the Evolution and genetics of Sexuality*, University of California Press
7. Hamilton, W.D. (1980) **Sex versus non-sex versus parasite**, *Oikos* 35, 282-290
8. Hamilton, W.D., Henderson, P.A. and Moran, N. (1980) **Fluctuation of environment and coevolved antagonist polymorphism as factors in the maintenance of sex**, in *Natural Selection and Social Behavior: Recent Research and Theory* (Alexander, R.D. and Tinkle, D.W., eds), pp. 363-381, Chiron Press
9. Hamilton, W.D. (1982) **Pathogens as causes of genetic diversity in their host populations**, in *Population Biology of Infectious Diseases* (Anderson, R.M. and May, R.M., eds), pp. 269-296, Springer-Verlag
10. LaMunyon, C.W. and Ward, S. (1997) **Increased competitiveness of nematode sperm bearing the male X chromosome**, *Proc. Natl. Acad. Sci. U.S.A.* 94, 185-189
11. Brenner, S. (1974) **The genetics of Caenorhabditis elegans**, *Genetics* 77, 71-94
12. Kenyon, C. (1988) **The nematode Caenorhabditis elegans**, *Science* 240, 1448-1453
13. Ward, S. and Caurel, J.S. (1979) **Fertilization and sperm competition in the nematode Caenorhabditis**

elegans, *Dev. Biol* 73, 304-321

14. Crouse, H. (1960) **Sex determination in Sciara**, *Genetics* 45, 1429-1443

15. Chandra, H.S. and Brown, S.W. (1975) **Chromosome imprinting and the mammalian X chromosome**, *Nature* 253, 165-168

16. Lyon, M.F. (1995) **X chromosome inactivation and imprinting**, in *Genomic Imprinting: Causes and Consequences* (Ohlsson, R., Hall, K. and Ritzen, M., eds), pp. 129-141, Cambridge University Press

17. Nur, U. *et al.* (1988) **A 'selfish' B chromosome that enhances its transmission by eliminating the paternal genome**, *Science* 240, 512-514

18. Gadagkar, R. (1997) *Survival Strategies: Cooperation and Conflict in Animal Societies*, Harvard University Press

19. Haig, D. (1997) **The social gene**, in *Behavioural Ecology: an Evolutionary Approach* (4th edn) (Krebs, J.R. and Davies, N.B., eds), pp. 284-304, Blackwell

＊ ＊ ＊ ＊ ＊ ＊ ＊ ＊ ＊ ＊ ＊ ＊ ＊ ＊ ＊ ＊ ＊ ＊ ＊

Questions:

1. What are three advantages to sexual reproduction?

2. Why can the X chromosome in the *C. briggsae* be described as a selfish chromosome?

3. Describe what a haplodiploid life cycle.

Answers are at the end of the book.

- 23 -

The human papillomavirus (HPV) appears to exploit a weakness in one of the body's star cancer fighters — the p53 protein. The less effective defense that results allows the virus to disrupt cell growth, and that change sometimes develops into cervical, vluval, penile, and perianal cancers. It has been known that viral oncogenes are pivotal in the development of cancer. However, more than just their expression is required before a malignant change in the cell occurs. A group of British and Canadian researchers have discovered one form of modification in specific cellular genes taht appear to predispose to HPV-associated cervical cancer.

Interestingly, the composition of the p53 protein can vary, depending on the p53 gene that encodes it. In some people, the gene encodes a protein with an amino acid called arginine in a specific location. In others, the protein has proline in that spot. Some people have one copy of each of the variants of the p53 gene. This study explores the significance that this variant may have in the susceptibility to HPV-associated cervical cancer.

Papillomavirus and p53

by Harald zur Hausen

Nature, May 21, 1998

Certain types of human papillomavirus (HPV) are linked with most cases of cervical cancer, and also with vulval, penile and perianal cancers[1,2]. Although viral oncogenes are obviously pivotal in the development of cancer, merely expressing them is not enough — either for immortalization of cultured human cells, or for malignant conversion. Rather, additional modifications in specific cellular genes are needed, and, on page 229 of this issue, Storey and colleagues[3] describe one such modification that predisposes to HPV-linked cervical cancer.

Almost ten years ago, a fresh perspective was brought on the possible mechanisms by which papillomaravirus contribute to cancer, when two HPV oncoproteins, E6 and E7, were shown to interact with two cellular proteins — p53 and retinoblastoma, respectively[4,5]. Both p53 and retinoblastoma had previously been shown to interact with oncoproteins of other DNA tumour viruses, including SV40 and some adenoviruses. After interaction with E6, p53 was found to be degraded[6], and because p53 is one of the most important cellular proteins in guarding repair processes and maintaining chromosomal stability, this could explain why mutational changes are observed in human cells that are immortalized in culture by HPV.

In human populations, the *p53* gene is polymorphic at amino acid 72 of the protein that it encodes — that is, p53 may contain either a proline or an arginine residue at this position. So far, no correlation has been made between either of these forms and specific human tumours (with the possible exception of lung cancer in nonsmokers). But Storey *et al.* now reveal that the arginine form of p53 is more susceptible to degradation by the HPV E6 protein than is the proline form. Moreover, patients with HPV-associated cervical cancer are much more likely to contain the arginine form of p53 compared with the rest of the population. The authors concluded that patients with two copies of the arginine form have a sevenfold higher risk of developing cervical cancer than people with the proline form. Interestingly, a high percentage of skin squamous-cell carcinomas have been reported[7,8] to contain HPV DNA, and these show an even greater prevalence of the arginine p53 modification.

If the arginine form of p53 binds more effectively to E6, it will be degraded (and hence inactivated) more rapidly, leading to increased mutation rates and chromosomal instability. Conversely, in the presence of E6, more functional activity should be maintained with the proline form of p53. Residual p53 activity has been shown in some cervical carcinoma cell lines that contain HPV, and it would be interesting to see whether these lines contain the arginine or the proline form of p53.

The results of Storey and colleagues support the theory that cellular genes must be modified for HPV-linked carcinogenesis to occur[9]. There is likely to be a gradual accumulation of specific cellular changes during malignant progression, and evidence for this includes the long latency for tumour development after primary infection, the observed monoclonality of anogenital tumours that contain HPV, and the absence of tumour-specific modifications in the viral oncogenes. A pronounced mutator phenotype — probably mediated here by increased degradation of p53 — would facilitate accumulation of changes, increasing the risk and reducing the time required for malignant progression.

The involvement of genetic factors in HPV-linked carcinogenesis has been postulated in the past[10], but careful epidemiological studies have been missing until now. In part, this is due to the high prevalence of HPV in all populations studied so far. Now that the functional consequences of the binding of E6 and E7 to polymorphic cellular proteins have been identified, we should be able to study the molecular genetics of affected populations.

In the future, other cellular genes that actively impair viral oncoproteins or suppress their transcription in nontransformed proliferating cells will probably be identified — indeed, the first candidates are already emerging[2]. Failure or inactivation of these proteins will divert cells that carry HPV genomes down pathways to malignancy. Such discoveries will broaden our currently narrow perspective on the molecular genetics of HPV-linked cancers and although Storey et al. have made an interesting and unexpected start, HPV research is still full of surprises. ❑

References

1. IARC *The Evaluation of Carcinogenic Risks in Humans* Vol **64** (Human Papillomaviruses, IARC, Lyon, 1995).
2. zur Hausen, H. *Biochem. Biophys. Acta* **1288**, F55-F78 (1996).
3. Storey, A. *et al. Nature* **393**, 229-234 (1998).
4. Dyson, N., Levine, A.J., Münger, K. & Harlow, E. *Science* **243**, 934-937 (1989).
5. Werness, B.A., Levine, A.J. & Howley, P.M. *Science* **248**, 76-79 (1990).
6. Scheffner, M., Werness, B.A., Huibregtse, J.M., Levine, J.M. & Howley, P.M. *Cell* **63**, 1129-1136 (1990).
7. Shamanin, V. *et al. J. Natl Cancer Inst.* **88**, 802-811 (1996).
8. de Villiers, E.M. *Biomed. Pharmacother.* **52**, 26-33 (1998).
9. zur Hausen, H. *Lancet* **343**, 955-957 (1994).
10. zur Hausen, H. & de Villiers, E.M. *Annu. Rev. Microbiol.* **48**, 427-447 (1994).

* *

Questions:

1. What is p53's major role in cellular processes? *guarding repair process & maintaining chromosomal stability*

2. True or False? According to this study, having a proline residue at amino acid position 72 makes one more susceptibile to degradation by the HPV E6 protein.

3. What is the evidence that supports the theory that a gradual accumulation of specific cellular change needs to occur before malignancy results?

Answers are at the end of the book.

1) long latency for tumour development after 1° infection

2) observed monoclonality of anogenital tumours that contain HPV

3) absense of tumour specific modification in viral oncogenes

- 24 -

An increasing body of molecular information resulting from advances in basic research is being incorporated into clinical practice by medical genetics. The process by which these research advances progress from the laboratory to the bedside and their medical, social, and legal impact is a topic of intense current interest. Some health care professionals have claimed that new genetic information may lead to discrimination in insurance and employment; change the way courts allocate responsibility for injury and resultant damages; and be inappropriately interpreted by the medical profession. One of the most controversial genetic tests that will be commercially offered to women in the next few months is a test that detects genetic mutations that predispose to breast cancer. The complex medical, societal, psychological, financial, and employment issues that surround this testing need to be ironed out before this sensitive testing becomes incorporated into the medical management of high risk women by their health care providers.

In September of 1994, the breast-ovarian cancer susceptibility gene BRCA1, on the chromosome 17q21 was discovered, and screening for breast cancer susceptibility genes has been the center of focus for many organizations. Nevertheless, medical journals describe the daily discoveries of disease-causing genes for many of the 4000 known genetic conditions. It is very clear that the issue of genetic testing does not only apply to BRCA1, but rather to all of us. As science continues to make stunning progress in identifying and isolating genes, we will learn about the role genetics play in countless diseases, and gene testing will come to have an impact on all of our lives.

Genetic Testing: Out of the Bottle

by Orla Smith

Nature Medicine, June 6, 1996

Tests for detecting genetic mutations that predispose to breast cancer have followed hot on the heels of the 1994 discovery of the first breast cancer susceptibility gene, *BRCA1*. Women with a family history of breast cancer who test positive for a *BRCA1* mutation have an 85 percent lifetime risk of developing the disease (and a 50 percent lifetime risk of developing ovarian cancer). In April of this year, the Genetics and I.V.F. Institute in Fairfax, Virginia, started offering a commercial test that detects the specific *BRCA1* mutation present in one percent of Ashkenazi Jewish women. By the end of 1996, Myriad Genetics in Utah will start marketing a more comprehensive test that should detect all *BRCA1* mutations. Although

many physicians and patients herald these new tests as an important contribution to breast cancer prevention, they are also concerned that without protective legislation in place (now under way in the US Congress), such sensitive information could be used as a basis for discrimination by employers and health insurance companies.

Antonio Wolff, a medical oncologist at the Winship Cancer Center at Emory University in Atlanta, says "many women are enthusiastic about genetic testing for breast cancer but change their minds after discussing the implications with a physician or a genetic counselor." Wolff notes that, at a recent public forum on genetic testing in Atlanta, the biggest fear voiced by women attending was of discrimination by employers and insurance companies who might learn of a positive test result. These fears are not without foundation; numerous documented cases exist of insurance companies denying coverage based on detection of a gene mutation predisposing to disease.

Many legislators are determined to legislate against such discrimination. The Kennedy-Kassebaum bill in the Senate and a similar bill in the House of Representatives were recently passed and are now in conference. Both bills specifically address prevention of insurance discrimination based on pre-existing illness and, at the eleventh hour, were amended to prevent insurers from denying coverage on the basis of testing positive for a gene mutation predisposing to disease. Francis Collins, the director of the Human Genome Project at the National Institutes of Health, applauds the legislation and its amendment, saying "this is the first time that the term genetic information has been included in a congressional bill and is an important step forward." However, although both bills would prevent insurance companies from denying benefits to people with pre-existing conditions or genetic predisposition to disease, neither bill prevents insurers from charging exorbitant rates to such individuals.

Although legislation is moving toward preventing discrimination based on a person's genetic composition, many point out that mutations predisposing to disease have always existed, but it is only now that we have the tools to detect them. According to David Sidransky, associate professor of oncology at the Johns Hopkins Medical Institutions, "we need to lift the mystique surrounding genetic predisposition. Doctors test for disease risk all the time — we get our cholesterol levels tested at the shopping mall and think nothing of it. I think there is a philosophical issue here — people see genes and mutations as predestined and unalterable whereas blood pressure and cholesterol are perceived as characteristics that can be changed." Collins points out "there are no perfect specimens. More and more genetic tests will become available and everyone will end up having a gene mutation predisposing to some disease." Although insurance

companies balk at the notion of providing coverage for an individual with a mutation predisposing to disease, a positive test can enable patients to take measures such as increased surveillance or prophylactic surgery to prevent future disease, thus saving insurance companies money in the long run.

Fear of discrimination is not the only troubling issue raised by genetic testing. "We don't know what the risk is for women without a family history of breast cancer who test positive for one of the known *BRCA1* mutations," says Neil Holtzman, professor of pediatrics at the Johns Hopkins Medical Institutions. "Among women with a family history of breast cancer who test positive for a *BRCA1* mutation, approximately 15% will not develop the disease. This percentage may be higher in women without a family history who test positive." This view is echoed by Paul Billings, a medical geneticist at the Palo Alto Veterans' Administration Health Care System and Stanford University, who says "the story is far more complex than anyone admits. Additional modifying genes and environmental factors may alter risk." Billings believes that testing for *BRCA1* mutations in women from high risk families, who have at least two or more close relatives with breast cancer, within a framework of intensive counseling, "is appropriate," although he doesn't advocate testing for the general population "at this time."

The American Society of Clinical Oncology, which published guidelines for genetic testing in the May issue of the *Journal of Clinical Oncology,* also recommends that only individuals with a strong family history of cancer or very early age of onset of the disease should be tested. Says Peter Meldrum, the president and CEO of Myriad Genetics, who cofounded the company in 1991 with Mark Skolnick (a codiscoverer of the *BRCA1* gene), "we will be offering our test to women at high risk of developing breast cancer." Although Meldrum says that Myriad Genetics will not be offering prenatal testing or testing of minors, he admits that if a woman without a family history of the disease wants to be tested — with the support of her physician — she will not be denied that opportunity. "We cannot be paternalistic," he says.

Most parties agree that genetic testing for breast cancer needs to be performed within academic centers using Institutional Review Board-approved protocols to assess the sensitivity, specificity and predictive value of the *BRCA1* tests. Ensuring that all commercial centers offering genetic testing follow established protocols might prove difficult, because they do not need the approval of the US Food and Drug Administration. Collins sees the solution in a massive expansion of the number of research protocols at academic centers in which women could enroll to obtain the *BRCA1* test while also receiving essential education and counseling. Meldrum emphasizes that women will not receive the Myriad Genetics test unless they sign a consent form stating that they have read and understood the educational material provided and have also received counseling. "Educating not only patients but health care providers about the benefits and limitations of testing for *BRCA1* mutations is essential," says Wolff.

Women testing positive for a *BRCA1* mutation have several choices in an attempt to prevent breast cancer. However, "there are little data indicating which is the best option for women with a positive test result," says Holtzman, "Bilateral mastectomy, frequent monitoring, prophylactic chemotherapy or mammography screening at a younger age need to be compared in a systematic manner, including randomized clinical trials."

"Yet ah! Why should they know their fate? . . . . where ignorance is bliss, 'tis folly to be wise." The prophetic and oft' misquoted words of Thomas Gray, the 18th-century American poet, hold part of the answer to the current dilemma. Is genetic predisposition a situation where ignorance is bliss? The advent of genetic testing has now made "ignorance" an option, but an option that, paradoxically, only the well-informed woman can choose. ❏

* * * * * * * * * * * * * * * * * * * *

Questions:

1. What are *BRCA1* alterations carriers at risk for?

2. Who would be most appropriate for *BRCA1* testing?

3. What is the biggest fear voiced by women who are considering genetic testing for *BRCA1*?

Answers are at the end of the book.

Part Seven

The Genetics
of
Evolution

- 25 -

In eukaryotes and prokayrotes, DNA serves as the molecule storing genetic information for the cell. In viruses, either DNA or RNA serves this function. DNA and RNA are the two types of nucleic acids found in organisms. Nucleic acids, along with carbohydrates, lipids and proteins make up the four major classes of organic biomolecules that characterize life on earth. DNA stores genetic information in the form of a code, a genetic code. This code, made up of four bases, specifies the amino acids or the chemical building blocks of proteins. The coded message in the DNA is transcribed and translated into a protein molecule by RNA. RNA is the workhorse of the genetic world, transcribing the coded instructions of DNA and assembling amino acids into proteins. Biologists have speculated about the nature of the first life forms and how their existence led to modern life. Some biologists think that the earliest life forms received their genetic instructions through RNA molecules. This "RNA world" may have been the crucial initiating step of life itself.

Let There Be Life

by Phil Cohen

New Scientist, July 6, 1996

What magic ingredients transformed a seething broth of chemicals into the first living organisms? Phil Cohen describes some new twists in the search for the bare necessities of life.

At 3•5 billion years old, fossilised bacteria are the earliest evidence of life on Earth, and yet these relics, with names like *Chroococcaceae* and *Oscillatoriaceae*, are identical to the sophisticated modern cyanobacteria that cover the globe from Antarctica to the Sahara. Evidence of any simpler incarnation fried in the intense heat of the young Earth before conditions were favourable for fossilising its remains.

In the absence of any rock-solid evidence, biologists have been free to speculate about the nature of the mysterious fledgling life form that came into existence some 4 billion years ago, and from which every plant, animal and microbe alive today eventually descended. All agree that early life, by definition, must have been capable of replicating and evolving. To do these things, most biologists have assumed that the ancestral life form needed a rudimentary instruction manual — a set of primitive genes — that was copied and passed from generation to generation.

In the past year or so, a majority view has emerged on which molecules first acquired these abilities and so sparked life on the planet Earth. Buoyed by some spectacular breakthroughs, most biologists are now convinced that life began when molecules called RNA took on the tasks that genes and proteins perform in today's sophisticated cells. In the once controversial "RNA world" theory, the chance production of largish RNA molecules was the crucial and committing step in the emergence of life itself. For many, this has become the only acceptable version of events.

But just when it looked safe to carve the RNA world theory in stone, its opponents are staging a spirited counterattack. Scientist in this second group don't agree on the details of their alternative visions, but they all make a claim that seems almost blasphemous in the era of molecular biology: far from being the first spark of life, they say, the RNA instruction manual was a mere evolutionary afterthought that helped fan its flames. What is more, they claim that the evidence proving their case will be in by the end of the decade.

A Tricky Problem

All modern life forms, be they germs, geraniums or Germans, have genes. The genes are made of DNA, which is made up of nucleotides; it is the sequence of these subunits that encodes the cell's instruction manual. The DNA is translated into RNA (also made up of nucleotides) which provides the blueprint for protein construction. The proteins, in turn, do all the metabolic grunt work, such as catalysing the chemical cycles that capture energy for the cell. They are also needed to translate DNA into RNA, and to make DNA copies to pass to the daughter cells. In other words, proteins, DNA and RNA are all essential for life as we know it.

For decades, this *ménage à trois* was the undoing of many a biologist trying to come up with believable scenarios for how life first appeared. Take away any one of the three and life grinds to a halt. But coming up with a plausible story for how DNA, RNA and proteins suddenly popped into existence simultaneously on a lifeless planet was just as tricky.

The first chink in this intellectual impasse appeared in the 1980s. Then Tom Cech at the University of Colorado and Sydney Altman at Yale University discovered that two naturally occurring RNA molecules sped up a reaction that snipped out regions of their own nucleotide sequence. RNA, it turned out, had some catalytic muscle of its own. The catalytic RNAs became known as ribozymes.

Theoreticians jumped on this discovery, envisaging a long ago world in which RNA ruled the planet. First, by virtue of its ability to act as a template for new RNA molecules, RNA was perfect for storing and passing on information. Second, by virtue of its ability to snap bonds between atoms, RNA was also a catalyst. Most crucial to the theory's credibility, the scientists proposed that RNA once catalysed the creation of fresh RNA molecules from their nucleotide building blocks.

Eventually, the free-wheeling RNA molecules would have acquired membranes and taken on additional catalytic tasks needed to run a primitive cell. But RNA's reign did not last. Under the pressure of natural selection, the proteins, which are better catalysts than RNA, and the DNA, which is less susceptible to chemical degradation, staged a cellular coup d'état, relegating RNA to its present role as a DNA-protein go-between.

Not surprisingly perhaps, those inclined to scepticism argued that it was too great a leap from showing that two RNA molecules partook in a bit of self mutilation in a test tube, to claiming that RNA was capable of running a cell single-handed and triggering the emergence of life on Earth.

Jack Szostak, a biochemist at Massachusetts General Hospital in Boston, set out to prove the sceptics wrong. He reasoned that the first RNA molecules on the prebiotic Earth were assembled randomly from nucleotides dissolved in rock pools. Among the trillions of short RNA molecules, there would have been one or two that could copy themselves — an ability that soon made them the dominant RNA on the planet.

To mimic this in the lab, Szostak and his colleagues took between 100 and 1000 trillion different RNA molecules, each around 200 nucleotides long, and tested their ability to perform one of the simplest catalytic tasks possible: cleaving another RNA molecule. They then carried out the lab equivalent of natural selection. They plucked out the few successful candidates and made millions of copies of them using protein enzymes. Then they mutated those RNAs, tested them again, replicated them again, and so on to "evolve" some ultra-effective new RNA-snipping ribozymes.

In the past few months, David Bartel, a biochemist at the Whitehead Institute for Biomedical Research near Boston and a former member of Szostak's team, has gone one better. He has evolved RNAs that are as efficient as some modern protein enzymes. The problem with most ribozymes is that they are as likely to snip an RNA molecule apart as stitch one together, which makes copying a molecule fifty nucleotides long (the minimum size necessary to catalyse a chemical reaction) a Sisyphean task. Bartel's new ribozymes, on the other hand, can stitch small pieces of RNA together without breaking larger molecules apart. What is more, his ribozymes use high-energy triphosphate bonds similar to ATP as their fuel, speeding the reaction up several million-fold.

"We've got ribozymes doing the right kind of chemistry to copy long molecules," says Szostak. "We haven't achieved self-replication from single nucleotides yet, but it is definitely within sight."

Electricity and Hot Air

Still, for the RNA world to have worked, it would have needed a supply of adenine, cytosine, guanine and uracil, the nucleic acid bases that, along with sugar and phosphate, make up nucleotides. Back in the 1950s, Stanley Miller, a 23-year-old doctoral student at the University of Chicago, announced that he had made amino acids, the pieces that click together to make proteins, with little more than a stuttering spark of electricity shot through hot air circulating in some glass tubing. The discovery was hailed as the first evidence that a lifeless planet could have spat out any of the raw materials needed for carbon-based life.

Miller's spark was a stand-in for primeval lightning, and the hot air, containing ammonia, hydrogen, water vapour and methane, was meant to mimic Earth's atmosphere 4 billion years ago. Besides creating amino acids, other researchers quickly demonstrated that the rich organic gook spewed out by Miller's decidedly non-biological combination also harboured chemical reactions that created huge amounts of adenine and guanine.

Cytosine and uracil, however, remained elusive. For this reason, and others, Miller's experiment did not convince everyone. Many atmospheric scientists argued that, unlike Miller's experimental setup, the incipient Earth was hydrogen-starved and entirely unsuitable for organic synthesis outside of a few havens, such as deep-ocean vents. This glitch led to the proposal of an alternative — to some fanciful — theory: that the organic building blocks came from outer space.

For much of his career, Jeffrey Bada, a geochemist at Scripps Institute of Oceanography, had argued that this was impossible. But a few months ago, Bada's own research forced him to change his mind. He found evidence that "mother lodes" of buckyballs have been delivered intact to Earth from outside the Solar System. Bada and his colleague Luann Becker made their find at Sudbury, Ontario, where a meteoroid the size of Mount Everest crashed 2 billion years ago. At first, Bada assumed that the buckyballs, football-shaped molecules made up of carbon atoms, had formed from vaporised carbon at the time of the impact. Then he discovered that they were loaded with helium, an element that has always been rare on Earth, but is abundant in interstellar space. What is more, the single impact site contained about 1 million tonnes of extraterrestrial buckyballs. If complex buckyballs could fall to Earth without being burnt up, so could complex organic molecules. "This blew our minds," says Bada. "We never expected it to be possible."

And while Bada's conversion was taking place, Miller, now at the University of California, San Diego, had not given up on the idea that the primeval organic slime — wherever it came from — could have spawned the missing nucleic acid bases, cytosine and uracil. Last summer, 43 years after his original experiment, he and his student Michael Robertson discovered a way for the primordial pond to make them by the bucketload. The secret ingredient was urea. Although urea is produced in Miller's original experimental setup, it never reaches a high enough concentration. But when he added more of the chemical, it reacted with cyanoacetaldehyde (another byproduct of the spark and hot air) churning out vast amounts of the two bases. Miller argues that urea would have reached high enough concentrations as shallow pools of water on the Earth's surface evaporated — the "drying lagoon hypothesis."

And in the last few months, another gap in the RNA world theory has been plugged. "The real question," says Jim Ferris, a chemist at Rensselaer

Polytechnic Institute in Troy, New York, "is how did we get from a prebiotic concoction to [the first] long pieces of RNA? What was the bridge to the RNA world?

In test-tube versions of the prebiotic world — as yet unblessed with protein enzymes or ribozymes — nucleotides link up, but only a few at a time. Once three or more have been connected, the RNA chain snaps — long before it has reached the magic length of fifty nucleotides needed to catalyse production of more RNAs.

In May, Ferris reported in *Nature* that he had found a means by which the first large chains could have been forged. When his team added montmorillonite, a positively charged clay that they think was plentiful on the young Earth, to a solution of negatively charged adenine nucleotides, it spawned RNA 10-15 nucleotides long. If these chains, which cling to the surface of the clay, were then repeatedly "fed" more nucleotides by washing them with the solution, they grew up to 55 nucleotides long.

The clay gets RNA off the hook of having to take on the tasks of information storage and catalysis in one fell swoop, says Ferris. It would catalyse RNA synthesis, stocking pools with a large range of RNA strands that, as Szostak and others have shown, would evolve a catalytic capacity of their own. In theory, an RNA catalyst would be born that could trigger its own replication from single nucleotides.

And with all the new evidence that is now available the apostles of the RNA world believe that their theory should be taken, if not as gospel, then as the nearest thing to truth that the science of the origins of life has to offer.

Not everyone agrees.

Power Shortage

Evolutionary biologist Carl Woese of the University of Illinois says the genetic evidence contradicts the RNA world theory. And if that weren't bad enough, he also argues that the RNA world scenario is fatally flawed because it fails to explain where the energy came from to fuel the production of the first RNA molecules, or the copies that would be needed to keep the whole thing going.

In test-tube RNA worlds, the elongating RNA molecules are fed artificially "activated" nucleotides, boosted with their own tri-phosphate bond to ensure that they come with an energy supply. In nature, such molecules only exist inside cells, and they have never been created in a Miller-type experiment. "The RNA world advocates view the soup as a battery, charged up and ready to go," Woese complains. On the primordial Earth, that energy had to come from somewhere, and it had to be coupled to production, or else it would quickly disappear into the ether.

In Woese's view, the critical step that ultimately spawned life was not a few stray RNA molecules, but the emergence of a biochemical machine that transformed energy into a form that was instantly available for the production of organic molecules.

The Energy Machine

Günter Wächtershäuser, an organic chemist at the University of Regensberg in Germany has suggested just such a machine. According to his picture, iron and sulphur in the primordial mix combined to form iron pyrites. Short, negatively charged organic molecules then stuck to its positively charged surface and "fed" off the energy liberated as more iron and sulphur reacted, creating longer organic molecules. The negatively charged surfaces of these molecules would attract more positively-charged pyrite, and the cycle would continue.

And by Wächtershäuser's reckoning, this energy-trapping cycle could easily have evolved into life forms that now exist — as chance ensured that one of the growing organic molecules was eventually of the right composition to catalyse its own synthesis. Ultimately, cycles of organic molecules would evolve that could trap their own energy — at which point they could do away with the inorganic energy cycle.

According to Woese, Wächtershäuser's theory and the RNA world theory are all testable, if only you know where to look for clues. The physical record of Earth's earliest life forms may have been erased, he says, but their "echoes, carried all the way through from precellular times" remain encoded in the genes of modern organisms.

Six years ago, Woese, with Otto Kandler of the University of Munich and others, used those clues to transform our understanding of recent evolution. By using the mutation rates of genes as their guide, they pruned the tree of life, which traces how different species evolved, from five main sections to just three.

Woese says that a similar type of genetic analysis now shows that, contrary to the view of RNA world advocates, replication of RNA appears to have been a late development in evolution, and not its starting point. If RNA molecules had been responsible for the emergence of life, then the ancestral cell — which was supposedly descended from the initial RNA life forms, and the ancestor of all current life forms — would have had a sophisticated machinery for copying RNA. The genes encoding that machinery would have been subjected to selection pressures from the get-go, and so should be present in every modern organism in a relatively unaltered state. But, says Woese, when biologists look at these genes, species from the three branches of the tree of life have little in common. That shows, says Woese, that the machinery needed to copy RNA was a work in progress in the common ancestor cell, and that subsequent evolution on the three branches of the tree solved its inefficiencies in very different ways.

In short, RNA replication could not have been the trigger for the emergence of life. "Only its mere essence was there at the time of the common ancestor," Woese says.

And, he warns, "we're only beginning to unlock the secrets of the common ancestor." Comparisons of genes may soon reveal the identity of the first energy-producing metabolic cycle, he says. Assuming, for a moment, that the metabolic cycle was the initial life form, then when the first genes appeared they would have been co-opted into ratcheting up the efficiency of the metabolic cycle by producing enzymes to catalyse each step. These genes would then have been subjected to selection pressures for longer than any others, and should be present in all modern organisms in a similar state.

Until recently, an all-out search for this first metabolic cycle has been impossible, because only bits and bobs of DNA sequence were available from a few organisms. But genome projects are gathering momentum, spewing out complete sequences of organisms' every gene faster than the scientists can analyse them. This month, Woese and his colleagues plan to be the first to publish the sequence of an archaebacterium, *Methanococcus jannaschii*, a resident of boiling, deep-ocean vents. Woese predicts that 100 whole genome sequences will be in the databases by the end of the decade. Enough, perhaps, to finally track down the primordial energy cycle.

Woese and Wächtershäuser may be ruffling the feathers of RNA world enthusiasts by suggesting that an energy producing metabolic cycle, not RNA, triggered life on Earth. But Stuart Kauffman, a theoretical biologist at the Santa Fe Institute in New Mexico is leaving them speechless by suggesting that life forms may exist that have no need of RNA or DNA or any other "aperiodic solid". What is more, he says, the emergence of life wasn't some chance event, but something that was bound to happen under the conditions of the primitive Earth.

Out of Chaos

Kauffman argues that the emergence of life on Earth is not the success story of a single type of molecule, such as RNA, slowly evolving to take on the catalytic burden of self-replication. In his view, the process was far more democratic. According to complexity theory, when a system reaches some critical level of complexity, whether it is made of stocks and shares or molecules, it naturally generates a degree of complex order. Likewise, he says, the mundane mix of nucleotides, lipids and amino acids that made up the primordial soup would in one magic instant have become an integrated system as the natural consequence of being part of a chaotic and complex mess.

Under such conditions, he says, self-replicating, "life-like" order is not a chance occurrence, it's a dead cert. In Kauffman's view, the modern *ménage à trois* of protein, RNA and DNA is not a conundrum, but a natural consequence of how life began.

He has demonstrated his theory using a computer model of the primordial stew. This shows that when a group of molecules — computer equivalents of simple organics with a few rudimentary catalytic skills — reach a critical level of diversity they spontaneously form an "autocatalytic set": a molecular cooperative that replicates as a group and evolves to create ever more complicated members. In other words, an autocatalytic set is a life form. What is more, says Kauffman, any sufficiently diverse mix — whether it is of carbon compounds or particles in an intergalactic dust cloud — will form autocatalytic sets, live, and evolve.

True, says Kauffman, RNA and DNA are part of all life today, but they arose as an accessory to an already flourishing ancestral autocatalytic set. Before genes existed, natural selection exerted its forces on the autocatalytic sets, ensuring that they were not biological dead-ends, but living systems capable of evolving to best suit their environment.

But many bench biologists scorn such ideas as cyberfantasy. "It's a pretty thought," says Gerald Joyce, who studies test-tube evolution at the Scripps Research Institute in San Diego. "But to be convinced, I need to see this autocatalytic gemish." And there's the rub. To prove Kauffman's theory you would need to analyse the contents of a pot in which percolated billions of different organic molecules, identify the autocatalytic entities and isolate them, and put them through their self-replicating cycles. Such an experiment stretches the bounds of what is technically feasible.

After years of trying to persuade the RNA world enthusiasts of the errors of their ways, however, Kauffman says he has gathered allies in biochemistry (he refuses to name names) who are willing to take on that task. He expects results in two to three years.

But in the short-term at least, most biologists say that the RNA world theory will prevail. Not unnaturally that worries those in opposition such as Woese, Kauffman, and Wächtershäuser.

"RNA chauvinism dominates the textbooks," says Gary Olsen, Woese's colleague at the University of Illinois. And that's a mistake, he warns, because the RNA world "as a theory it is only partly proven. The rest is speculative optimism." ❑

* * * * * * * * * * * * * * * * * * * *

Questions:

1. What is the problem with most ribozymes?

2. What two nucleic acid bases were created as a result of Miller's experiment?

3. Describe the complexity theory as it relates to the emergence of life

Answers are at the end of the book.

- 26 -

Understanding the life-styles of extinct animals is a daunting task that has been made easier by findings of molecular biologists. They have discovered a chemical trick for extracting ancient DNA from desiccated dung left by ice age animals in the Pleistocene era. This analysis could be performed only after the DNA was first freed from sugar-derived condensation products through an agent that cleaves the sugar links. Analysis of the amplified DNA, which ranged in size from 153 to 273 base pairs, indicated that the sloth had ingested representatives of six families and two orders of plants. The new procedure could help in a variety of biological endeavors, from studies of extinct saber-tooth cats to analyses of endangered living populations.

Molecular Coproscopy: Dung and Diet of the Extinct Ground Sloth *Nothrotheriops shastensis*

by Hendrik N. Poinar, *et al.*

Science, July 17, 1998

DNA from excrements can be amplified by means of the polymerase chain reaction. However, this has not been possible with ancient feces. Cross-links between reducing sugars and amino groups were shown to exist in a Pleistocene coprolite from Gypsum Cave, Nevada. A chemical agent, N-phenacylthiazolium bromide, that cleaves such cross-links made it possible to amplify DNA sequences. Analyses of these DNA sequences showed that the coprolite is derived from an extinct sloth, presumably the Shasta ground sloth Nothrotheriops shastensis. *Plant DNA sequences from seven groups of plants were identified in the coprolite. The plant assemblage that formed part of the sloth's diet exists today at elevations about 800 meters higher than the cave.*

The polymerase chain reaction (PCR) has opened up new possibilities in ecology, archaeology, and paleontology by making it possible to retrieve DNA sequences from several previously untapped sources (1). One such source is feces, from which amplified DNA sequences allow identification of the species or individual from which the droppings originated as well as aspects of the diet and parasitic load of the animal (2). Large amounts of ancient feces (coprolites) can be found in certain dry caves or rock shelters — for example, in the southwestern United States (3). To investigate whether it may be possible to amplify DNA from such material, we have analyzed coprolites from Gypsum Cave about 30 km. east of Las Vegas, Nevada. These have been attributed to the Shasta ground sloth, *Nothrotheriops shastensis* (4), which became extinct about 11,000 years ago (5). Initial experiments showed that extraction protocols developed for contemporary fecal material did not yield amplifiable DNA from the Gypsum Cave coprolites (6). Therefore, to investigate the general state of macromolecular preservation in the coprolites, we performed chemical analyses.

Several samples were removed from a large fecal bolus. One was dated by accelerator mass spectrometry to $19,875 \pm 215$ years (Ua11835). Another sample was subjected to pyrolysis gas chromatography — mass spectrometry (Py-GC-MS). The pyrolysate was dominated by products related to polysaccharides (cellulose and hemicellulose) and lignin. Minor pyrolysis products derived from proteins were also observed. Although some oxidation of lignin is evident, the relative abundance of polysaccharide products, and especially the hemicellulose derivative, suggest a relatively high degree of chemical preservation of ligno-cellulose (7). The large amount of syringol derivatives demonstrates that angiosperm lignin dominates in the coprolite, whereas the presence of vinylphenol can be attributed to monocotyledonous lignin.

The volatile components of another sample of the bolus were analyzed by head space GC-MS (8). Alkylpyrazines, furanones, and furaldehydes, all products of the condensation of carbonyl groups of reducing sugars with primary amines (the Maillard reaction) (9) were detected. Specifically, the alkylpyrazines are formed through condensation of dicarbonyl compounds with amino acids (9, 10), and the furan products are indicative of advanced stages of the Maillard reaction, which may result in extensive cross-linking of macromolecules (9) such as proteins (11) and nucleic acids (12). Maillard Products have been suggested to be a prominent component in ancient DNA extracts (13) and were recently found in ancient Egyptian plant remains (8).

Because these analyses indicated the presence of Maillard products, N-phenacylthiazolium bromide (PTB), a reagent that cleaves glucose-derived protein cross-links (14), was added to the extraction mixtures to release DNA that might be trapped within sugar-derived condensation products. Extractions were done with and without the addition of PTB (15) and amplifications of a 153-base pair (bp) DNA segment of the mitochondrial 12S ribosomal RNA (rRNA) gene were attempted (16). From all extracts with added PTB, PCR products were observed, whereas no PCR products were observed in the absence of PTB. Furthermore, higher concentrations of PTB resulted in slightly stronger PCR products.

Four sets of primers (6, 17) were used to determine the nucleotide sequence of a 572-bp segment of the 12S rRNA gene (16). The PCR products ranged in size from 153 to 273 bp and showed an inverse relationship of amplification strength to length typical of ancient DNA (18). The products were cloned and a minimum of 10 clones were sequenced (19). To ensure the authenticity of the sequence information to the greatest extent possible, a second extract was prepared (20) and the same four fragments were amplified, cloned, and sequenced. The sequence determined from the coprolite was aligned to homologous sequences from representatives of each family of extant edentates — two-toed and three-toed sloths, anteaters and armadillos, and one extinct ground sloth, *Mylodon darwinii*, previously sequenced from bones and soft tissue remains (6). The dung sequence differs at 41 to 55 positions from the extant tree sloths and the mylodont sloth at 88 positions from the other edentates. Phylogenetic analysis (21) confirm that the DNA sequence derived from the Gypsum Cave coprolite belongs to a sloth, presumably N. shastensis (Megatheriidae), whose bones were found in the cave (4).

To analyze the diet of the Gypsum Cave ground sloth, we treated DNA extracts with PTB and amplified a 183-bp fragment of the chloroplast gene encoding the large subunit of ribulose-bisphosphate carboxylase (rbcL) (16). We cloned the PCR product and sequenced the inserts of several clones. Ancient DNA contains chemical modifications that may cause substitutions in the amplified sequences, but such errors are unlikely to occur at the same position in sequences that are amplified independently (22). Therefore, we repeated the amplification and cloning twice with a different extract. We compared each unique insert with the approximately 2300 rbcL sequences deposited in the GenBank database (release 1.4.11; 24 November 1997) by means of the BLAST program (23) and noted the order and family of Gen Bank sequences displaying zero, one, and two differences from the clone. We classified clones with three or more differences as unassigned. We sequenced clones until 18 consecutive clones resulted in no new family assignments. In total, we sequenced 72 clones.

Fifteen clones were found to be identical to GenBank sequences from the families Capparaceae, Poaceae, Liliaceae, and Euphorbiaceae. Eleven clones were found to differ at one position from sequences derived from those four families. In addition, two clones differed at one position from Rubiaceae and Chenopodiaceae sequences. Finally, seven and nine clones, respectively, displayed two differences from sequences belonging to the families Capparaceae and Liliaceae, and two clones displayed two differences from sequences in the order Lamiales and Scrophulariales, which are closely related and sometimes regarded as one order (24). Whenever a group of related sequences was assigned to a taxon within a single family, we considered that family as the putative source of the clones. When a group of related clones matched species within different families belonging to one order, we considered that order as the putative source of the clones. Finally, when clones matched data bank sequences from different orders, we considered the plant as unassigned, most likely because no member of that order has yet been sequenced. With these criteria, 46 of 72 clones could be assigned to taxa in the database. Of the remaining 26 clones, most are closely related to the groups of sequences described above and carry substitutions observed in only one clone or in a few clones from one and the same amplification. These are likely due to damage in the template. Three clones appear to represent recombination products between other sequences in the sample and are therefore likely to be artifacts created during the PCR — a phenomenon previously observed in ancient DNA (25). Thus, analysis of the rbcL sequences suggests that the coprolite contains the remains of at least seven plants ingested by the Pleistocene ground sloth. Five of these can be assigned to an order or a family. These are, in order of frequency among the clones: Capers and mustards (Capparales), lilies and allies (Liliaceae including Agavaceae), grasses (Poaceae), borages and mints (Lamiales/Scrophulariales), and saltbushes (Chenopodiaceae). Currently there are two plants

that cannot be identified. To investigate whether other fragments of the *rbcL* gene yield different plant identifications, we amplified three additional *rbcL* gene fragments that partially overlap with the initial fragment and we sequenced a total of 116 clones. The three taxa that were observed in three or more clones in the first fragment (Capparales, Poaceae, Liliaceae) were also identified from these clones, whereas the taxa observed in only one or two clones were not seen. Furthermore, the families Vitaceae (grape), Hydrophyllaceae ("water leaf family"), and Malvaceae (mallows) were identified, each from single clones. Thus, whereas the sequencing of additional clones may reveal plants that are rare in the bolus, the identification of plants whose DNA dominates quantitatively in the coprolite appears to be achieved reliably with the 183-bp fragment. Nevertheless, it may be advisable to verify identification of the dominating plants by using an additional set of primers because a particular primer may select against a particular plant species as a result of substitutions in the primer sites.

Six families and two orders of plants are identified from the coprolite. Of these, Capparales and Liliaceae, make up 24 and 19%, respectively, of the 188 rbcL clones sequenced; the other five are rare in the bolus. It is interesting to note that all taxa identified on the basis of the DNA sequences are represented by contemporary genera in the Southwest. However, two taxa [Liliaceae (yucca and agave) and Vitaceae] do not occur in the vicinity of the cave today (elevation about 580m). Of these Liliaceae, most likely represented by *Yucca* spp. and *Agave* spp., is common in the sample and is the likely source of the monocot signal observed in the Py-GC-MS analysis. These genera, as well as all other taxa observed in the sample, are now common in high-elevation desert scrub (above about 1370 m) on the Spring Range, about 50 km west of Gypsum Cave, and on the Las Vegas Range, about 30 km north-northeast of the cave. At 19,875 ± 215 years before present (B.P.), this sample dates to the last glacial maximum and it is reasonable to assume that yucca, now found only at higher altitudes, would then have been common around the cave. The recovery of DNA attributable to the grape family (Vitaceae) is also of interest. The only representative of this family in the Mojave Desert is the wild grape (*Vitis*), an obligate hydrophile that occurs around springs and streams. The closest known paleosprings of relevant age are in Las Vegas Valley about 20 km to the southwest. Alternatively, Las Vegas Wash, located only 10 km south of the cave, may have experienced perennial flow during the glacial maximum. Thus, the ground sloth may have visited water sources a substantial distance from the cave.

A macroscopic examination of the same bolus identified five plant species, one of which was not seen by the molecular analysis. Conversely, four of the taxa identified from the DNA sequences were not seen in the morphological analyses. Of the plants seen by only one approach, most are rare in the sample and thus may have been missed because of stochastic effects. However, the notable exception is capers and mustard, which is the most frequent taxon among the clones sequenced (24%); yet it is not seen in the macroscopic analysis. This may be a result of the lack of macroscopically distinctive remains left in the bolus. It is doubtful that organs other than seeds of capers and mustard species would retain their morphological attributes after mastication and digestion. Thus, the molecular evidence may be able to detect plants that are difficult to identify morphologically. A further advantage of the molecular approach is that it uses defined criteria for identification of plants that can be criticized and improved, especially as the DNA sequence data banks increase in size. For example, the taxonomic identification would probably become more accurate if relevant species from the area were sequenced. An additional advantage of the molecular approach to scatology is that DNA sequences can be used to identify the defecating species and to study its phylogeny.

DNA amplification from the coprolite was possible only after sugar-derived cross-links had been resolved by PTB. To date, 18 coprolites that range in age between 10,900 and 31,410 B.P. and are derived from various extinct species have been extracted with and without PTB. Whereas none of these has yielded amplification products when PTB was not used, six (from Gypsum Cave, Rampart Cave, and Steven's Cave in the southwestern United States and from Ultima Esperanza Cave in southern Chile) have yielded animal DNA sequences when treated with PTB (26). If the chemical state of preservation of biomolecules in fossils is better understood, reagents such as PTB can hopefully be used to make additional types of samples available for DNA amplification. ❑

Reference and Notes
1. S. Pääbo, *Sci. Am.* **269**, 86 (Nov. 1993; _____, R.G. Higuchi, A.C. Wilson, *J. Biol. Chem.* **264**, 9709 (1989).
2. M. Höss, M. Kohn, F. Knauer, W. Schröder, S. Pääbo, *Nature* **359**, 199 (1992); M. Kohn, F. Knauer, A. Stoffella, W. Schröder, S. Pääbo, *Mol. Ecol.* **4**, 95 (1995); M. Kohn and R.K. Wayne, *Trends Ecol. Evol.* **12**, 223 (1997).
3. P. Martin, B. Sabels, D. Shutler Jr., *Am. J. Sci.* **259**, 102 (1961); R. Hansen, *Paleobiology* **4**, 302 (1978).
4. M.R. Harrington, *Gypsum Cave, Nevada, Southwest Museum Papers* (Southwest Museum, Los Angeles, CA, 1933), vol. 8; J.D. Laudermilk and P.A. Munz, *Carnegie Inst. Washington Publ.* **453**, 31 (1934).
5. A. Long and P.S. Martin, *Science* **186**, 638 (1974).
6. M. Höss, A. Dilling, A. Currant, S. Pääbo, *Proc. Natl. Acad. Sci. U.S.A.* **93**, 181 (1996).
7. B.A. Stankiewicz, M. Mastalerz, M.A. Kruge, P.F. van Bergen, A. Sadowska, *New Phytol.* **135**, 375 (1997); M.M. Mulder, J.B.M. Pureveen, J. Boon, A.T. Martiniez, *J. Anal. Appl. Pyrol.* **19**, 175 (1991); C. Saizjimenez and J.W. de Leeuw, *Org. Geochem.* **6**, 417 (1984); P.F. van Bergen *et al.*, *Geochim. Cosmochim. Acta* **58**, 3823 (1994).
8. R.P. Evershed *et al.*, *Science* **278**, 432 (1997).
9. J. Mauron, *Prog. Food Nutr. Sci.* **5**, 5 (1981).
10. H. Weenen, *et al.*, *Am. Chem. Soc. Symp. Ser.* **543**, 142 (1994).
11. R. Bucala, P. Model, A. Cerami, *Proc. Natl. Acad. Sci. U.S.A.* **81**, 105 (1984).
12. A. Lee and A. Cerami, *Mutat. Res.* **179**, 151 (1987); A. Papoulis, Y. Al-Abed, R. Bucala, *Biochemistry* **34**, 648 (1995).
13. S. Pääbo, in *PCR Protocols and Applications: A Laboratory Manual*, M.A. Innis, D.H. Gelfand, J.J. Sninsky, T.J. White, Eds. (Academic Press, San Diego, CA, 1990), pp. 159-166.
14. S. Vasan *et al.*, *Nature* **382**, 275 (1996).
15. PTB ($C_{11}H_{10}NOSBr$) was synthesized as described (14). The melting point of the product was 224° to 227°C (expected, 223° to 223.5°C) and its elemental composition was 4.88% N, 46.46% C, and 3.86% H (expected, 4.93% N, 46.49% C, and 3.55% S). The coprolite was teased apart and about 5 g was ground

under liquid nitrogen in a freezer/mill 6700 bone grinder (Spex Industries, Edison, NJ). To samples of 0.1 g of powder, 1 mi of buffer [as in S. Pääbo, *Proc. Natl. Acad. Sci. U.S.A.* **86**, 1939 (1989)] , and 20 ml of proteinase K (10 mg/ml) was added and the sample was incubated under agitation for 48 hours at 37°C. Two samples received 10 and 100 μl of a 0.1 M PTB solution in 10 mM sodium phosphate buffer (pH 7.4) before and another two after the organic extraction, whereas two received only the phosphate buffer. Samples were phenol extracted as described (20) and concentrated to about 200 ml with Centricon-30 microconcentrators (Amicon, Beverly, MA). All but one extraction were purified [as in M. Höss and S. Pääbo, *Nucleic Acids Res.* **21**, 3913 (1993)] except that only one 1.2 wash was performed and the silica pellet was washed twice with ice-cold New Wash (Bio 101, LaJolla, CA). Mock extractions with and without PTB were performed alongside all extractions. It may be noteworthy that eight extractions from Pleistocene bone samples and one 2300-year-old Egyptian animal mummy using PTB have to date failed to yield amplification products. It may be that Maillard products are particularly prevalent in coprolites.

16. PCR amplifications were performed [as in M. Höss and S. Pääbo, *Nucleic Acids Res.* **21**, 3913 (1993)] using the following primers for the 125 rRNA gene and the chloroplast *rbcL* gene: 12ss, 5'-AATTTCGTGCCAGC CACCGCGGTCA-3'; 12st, 5'-AAGCTGTTGCTAGTA GTACTCTGGC-3'; 12sa, 5'-CTGGGATTAGATACCC CACTAT-3'; 12so, 5'-GTCGATTATAGGACCAGGT TCCTCTA-3'; 12sd, 5'-TAAAGGACTTGGCGGTGC TTCAC-3'; 12sn, 5'-CCATTTCATAGGCTACACCT TGACC-3'; 12shp, 5'-GCACAATTATTACTCATAAG C-3'; 12sb, 5'-TGACTGCAGAGGGTGACGGGCGGT GTGT-3'; rbcL Z1, 5'-ATGTCACCACAAACAGAG ACTAAAGCAAGT-3'; rbcL 19, 5'-AGATTCCGCAG CCACTGCAGCCCCTGCTTC-3'; rbcL h1aR, 5'-GAGGAGTTACTCGGAATGCTGCC-3'; rbcL h1aF, 5'-GGCAGCATTCCGAGTAACTCCTC-3'; rbcL h2aR, 5'-CGTCCTTTGTAACGATCAAG-3'.

17. M. Höss, thesis, Ludwig-Maximilians University, Munich, Germany (1995).

18. O. Handt, M. Höss, M. Krings, S. Pääbo, *Experientia* **50**, 524 (1994).

19. DNA sequences for 12S rRNA are available at www.sciencemag.org/feature/data/981503.shl.

20. Determination of DNA sequences from two or more amplifications is necessary to detect polymerase errors in the first cycles of the PCR because the amplification may start from a single or a few DNA molecules [M. Krings *et al*, *Cell* **90**, 19 (1997)]. We consider reproduction of results in a second laboratory in order to exclude laboratory-specific contaminations to be necessary for human remains [O. Handt *et al., Science*

264, 1775 (1994); M. Krings *et al., Cell* **90**, 19 (1997)]. For animal remains, contaminations with human DNA are more easily detected and thus reproduction in different laboratories may be necessary only if extraordinary results are obtained. In this case, we believe that the sequence is authentic because it has been partially determined from five different boluses and found to be identical, and the phylogenetic analysis shows it to be related but not identical to a *Mylodon* sequence previously determined in our laboratory (6) and subsequently reproduced in another laboratory [P. Taylor, *Mol. Biol. Evol.* **13**, 2839 (1996)]. Furthermore, control extracts performed in parallel with the coprolite extractions yielded no amplification products and no work on these boluses had previously been performed in our laboratory. Finally, nuclear insertions are unlikely sourcers of the sequences because multiple primers yield the same sequence [M. Krings *et al., Cell* **90**, (1997)].

21. K. Strimmer and A. von Haeseler, *Mol. Biol. Evol.* **13**, 964 (1996).

22. M. Höss, P. Jaruga, T.H. Zastawny, M. Dizdaroglu, S. Pääbo, *Nucleic Acids Res.* **24**, 1304 (1996); O. Handt, M. Krings, R. Ward, S. Pääbo, *Am. J. Hum. Genet.* **59**, 368 (1996).

23. S.F. Altschul, W. Gish, W. Miller, E.W. Myers, D.J. Lipman, *J. Mol. Biol.* **215**, 403 (1990).

24. R.T. Thorne, *Bot. Rev.* **58**, 225 (1992).

25. S. Pääbo, D.M. Irwin, A.C. Wilson, *J. Biol. Chem.* **265**, 4718 (1990).

26. H.N. Poinar *et al.*, unpublished data.

27. B.A. Stankiewicz *et al., Science* **276**, 1541 (1997).

28. J.C. Hickman, Ed., *The Jepson Manual, Higher Plants of California* (Univ. of California Press, Berkeley, CA 1993).

29. W.G. Spaulding, in *Packrat Middens: The Last 40,000 Years of Biotic Change,* J.L. Betancourt, T.R. Van Devender, P.S. Martin, Eds. (Univ. of Arizona Press, Tucson, AZ, 1990), chap. 9.

30. We thank J. Bada, V. Bömer, D. Caccese, G. Eglinton, A. Greenwood, M. Höss, G. Poinar Jr., U. Schultheiss, W. Storch, L. Vigilant, and H. Zischler for help and discussions; the Natural History Museum of Los Angeles County, the Natural History Museum (London) (P. Andrews), and Northern Arizona University, Flagstaff (J. Mead), for coprolite samples; and the Deutsche Forschungsgemeinschaft (S.P.) for financial support. B.A.S., H.B., and R.P.E. thank J.F. Carter and A. Gledhill for technical assistance with the mass spectrometry and head space analysis, which were performed under Natural Environment Research Council grant GST/02/1027 to D.E.G. Briggs and R.P.E.

* * * * * * * * * * * * * * * * * * * *

Questions:

1. What can PCR on feces tell paleontologists about the species from which the dropping originated?

2. What is a coprolite?

3. What does N-phenacylthiazolium bromide do to DNA?

Answers are at the end of the book.

The Human Genome Diversity Project (HGDP) was proposed in 1991 to take a survey of the variation in human genome. The goal of HGDP is to find a more precise origin of world populations by integrating genetic knowledge, derived by applying the new techniques for studying genes, with knowledge of history, anthropology, and language. Some values of the HGDP is to help the world understand human history and identify genetic factors related to disease in some populations, and to help contribute to the elimination of racism. The blood samples taken by the HGDP will be immortalized for future study by transforming the samples into cell lines, which are capable of producing a large amount of duplicate DNA for study. The immortalized cell lines will be stored in various gene banks around the world and will be made available to qualified scientists interested in doing research on them.

Nicknamed the "Vampire Project," the HGD Project was formally adopted by the Human Genome Organization in January 1994. The Human Genome Project is a multinational, multibillion-dollar initiative by scientists, which seeks to sequence the DNA in the entire human genetic structure. The HGDP seeks to identify the genetic variations among populations. The intent of the HGDP is best described in this chilling description of the HGDP by Walter Bodmer, former president of the Human Genome Organization in the Book of Man: The Human Genome Project and the Quest to Discover Our Genetic Heritage.

> *"One of the most important components of the Human Genome Project will therefore involve the setting up of a type of rescue genetic archaeology. This will consist of the sampling of all the world's threatened peoples before their unique stock is lost or mixed with Western genes; this is one aim of the Human Genome Diversity Project, which has been set up to sample blood and tissues from indigenous peoples all over the world in order to create a unique genetic collection of the different human races."*

Creating this molecular Noah's Ark will be one of the most significant accomplishments of the Human Genome Project.

The Human Genome Diversity Project: Medical Benefits Versus Ethical Concerns

by Robert W. Wallace

Molecular Medicine Today, February 1998

By the year 2005 the entire human genome should have been sequenced and the genes identified. But the resulting genomic sequence, although a marvelous accomplishment, will be a composite of just a handful of individuals selected at random. The Human Genome Diversity Project was proposed as a means to overcome these limitations by obtaining genetic information from many diverse populations of the world. This would give medical geneticists a handle on the variations in susceptibility to disease among different populations, as well as being of anthropological value. But would such a project risk exploiting the indigenous populations involved?

Reprinted from *Molecular Medicine Today*, Vol. 4, pp. 59-62, by R.W. Wallace, "The Human Genome Diversity Project: Medical Benefits Versus Ethical Concerns," 1998, with permission from Elsevier Science.

The Human Genome Diversity Project (HGDP) is a proposal for an international collaborative effort to collect genetic material from a large number of indigenous peoples of the world and to make it available for systematic genetic studies. The project was first proposed in 1991 by Luigi Luca Cavalli-Sforza (Stanford University, CA, USA) and colleagues in a letter to the journal *Genomics*[1]. The original purpose of the project was to 'understand human diversity, both normal variation and that responsible for inherited diseases'. Its focus was to be the collection of nuclear and mitochondrial DNA from isolated populations that were likely to be linguistically and culturally distinct, and surrounded by geographic barriers.

In a series of meetings that followed the proposal, it was decided that the number of individuals to be studied would be between 10,000 and 100,000 from some 400-500 populations. For most of the study subjects, only small amounts of DNA would be prepared, stored and made available to qualified investigators from around the world. For approximately 10% of the sampled individuals, cells would be collected, transformed, and stored as frozen cell lines. This latter strategy, it was reasoned, would allow for the production of additional DNA should it be needed for future studies.

The original project envisioned that the collection of DNA would be supervised by regional committees, who would also collect a substantial amount of information regarding the people from whom the samples were collected. This information was to include: genealogical relationships among sampled individuals; birthplace and place of residence of the individual and of his/her parents; if applicable, clan of individual, father and mother; and birthplace, ethnic affiliation and clan of spouse (unless both husband and wife have been sampled separately).

Ethical guidelines were also agreed upon and included (1) obtaining informed consent before collecting samples, (2) respect for the privacy of individuals samples, (3) encouraging the participation of the population in the development of study designs and (4) keeping them fully informed of the results of the various studies.

The cost of the HGDP was estimated to be in the range of US $25-30 million — a mere one percent of the cost of the Human Genome Project. This might seem a gross underestimation when one considers that the Human Genome Project will only sequence one composite genome. However, much of the cost of the Human Genome Project is invested in developing the technology necessary to conduct massive-scale sequencing operations; by contrast, the costs of the HGDP are primarily for the collection of the genetic material, establishing a database of information on the study subjects and a system for the worldwide distribution of samples to researchers.

How could the HGDP benefit medicine?

Although its primary purpose is anthropological, many of the perceived benefits of the project are medical, because the HGDP would allow the assessment of the prevalence and distribution of disease genes in the populations of the world.

A better handle on genetic susceptibility to disease?

Sir Walter Bodmer, former president of the Human Genome Organization, believes that one of the most important medical outcomes of the HGDP will be a more detailed mapping of the location of genetic markers in 'normal' populations and their use in linkage disequilibrium studies. Such studies have already been used to identify some disease-causing genes, such as *CFTR*, the gene for cystic fibrosis. The approach is to use a set of gene markers to determine the degree of association of each marker with the disease of interest in the population. By using a wide range of markers with known locations, the location of the unknown gene responsible for the disease can be narrowed down to an ever-smaller region of the genome, eventually pinpointing its location. Bodmer points out that linkage disequilibrium studies have the potential to identify gene variations involved in complex, multifactorial genetic conditions, such as alterations of angiotensin or angiotensin-converting enzyme in cardiovascular disease, some forms of cancer, or variable drug responses. But the key to the use of the technique is to understand, in the 'normal' population, the frequency and distribution of whatever markers are being used. Such a study is necessary, says Bodmer, to allow one to say that the frequency of the marker in the disease is significantly different from that of the controls.

Henry Greely, Professor of Law at Stanford University (CA, USA), cites a number of other examples of possible medical benefits of studying the DNA of indigenous populations2. One is likely to be a better understanding of the high incidence of adult-onset diabetes through studies of narrowly defined populations such as the Pima tribe in Arizona, the Old Order Amish, the 'Gullah' Islanders in the American Southeast and some West Africans, all of whom suffer from an unusually high incidence of the disease. Studies have shown that diabetes within the Pima is 'strongly familial, and probably of genetic origin, although the precise nature of the gene or genes remains unknown'[3]. Greely also suggests that studies of specific populations such as the Old Order Amish might be useful in identifying the genetic component of conditions such as mental illness[4]. Studies of Jews of Eastern European origin may contribute to an understanding of the genetics of Tay-Sachs and Gaucher's disease and the 185delAG mutation associated with breast and ovarian cancer in Ashkenazi Jews[5].

The flip side: genes that protect from disease.

Greely also explains that some of the populations studied might possess genes that could be beneficial to the world population as a whole. Possible examples of such useful genes include those from the people of a village in Northern Italy who appear to be protected from some forms of heart disease, and prostitutes in Nairobi who appear to be immune to HIV infection[2,6].

Biotechnology and indigenous populations

Although the leaders of the HGDP consistently downplay the commercial aspects of the project, several biotechnology companies are already funding their own miniature versions of the HGDP in the search for useful genes that could lead to commercial products. One of these is Sequana Therapeutics Inc. (La Jolla, CA, USA) who, along with its major pharma partner Boehringer Ingelheim (Ingelheim, Germany), is prospecting for genes related to asthma from a variety of selected populations, including island populations in the south Atlantic, a Jewish community that was located in India for several thousand years, the Polynesian descendants that now occupy Easter Island, and large families in Brazil and China. These groups all have high incidences of asthma. Sequana hopes that an asthma gene can be identified, with the possibility that such a find will lead to new commercially viable therapeutic approaches for treatment[8]. Compared with the HGDP, the level of controversy over the efforts of these companies to collect genetic material has been modest. This could be because the companies have the funds to quietly go about the business of collecting genetic material without the public debate that accompanies the use of government research funds for such activity.

Other entrepreneurs have also been attracted by the potential commercial bonanza of DNA from indigenous peoples. Kari Stefansson, a Harvard geneticist and native of Iceland, has recently established a small company, deCode Genetics, Inc. (Reykjavik, Iceland), whose purpose is to collect DNA from the native population of Iceland to search for useful genes that will return a profit to the company and to the local people[7]. Stefansson is counting on the geographic isolation of the islanders, long-established patterns of maintaining medical records, selection pressure from an epidemic of the bubonic plague in the 1400s and severe famine brought on by a volcanic eruption in the 1700s, as well as the similar familial backgrounds, to have naturally selected for genes with commercial potential. Currently some 25 common diseases are under investigation at deCode, including multiple sclerosis, psoriasis, pre-eclampsia, inflammatory bowel disease, aortic aneurysm and alcoholism. The company is looking for commercial partners to further its efforts in mining the genome of the Icelandic people.

Stefansson also seems to have avoided much of the criticism focused on the HGDP, possibly because he is sampling a population of some 270,000 Icelandic citizens of which he is a member. In addition, deCode's approach to collecting samples goes to great lengths to maintain the privacy of the donor and only uses sample obtained from volunteers who have been fully informed and who have freely given their consent. Finally, in his agreements with pharmaceutical companies, Stefansson insists that any drugs developed as a result of the genetic information obtained from his sample population be supplied for use to the Icelandic population free of charge during the life of the patient.

Opposition to the HGDP
Reactions of indigenous peoples

Much to the surprise of the organizers of the HGDP, many of the populations of the world recoiled in horror at the prospect of having their DNA collected and studied. David Maybury-Lewis (Professor of Anthropology, Harvard University, Boston, MA, USA) explains that many indigenous peoples 'consider their land, the life forms on it and all aspects of their own persons — such as blood, hair and tissue, as well as DNA samples — to be sacred'[8] He further points out that such beliefs are often also held by non-indigenous people who are opposed to commercial ownership and exploitation of such a fundamental element of life as DNA.

Some populations are particularly concerned that useful DNA sequences found through the HGDP might be patented by US or European countries and exploited for commercial gain. RAFI (The Rural Advancement Foundation International), a non-profit, private organization in Montreal, Canada, has emerged as the champion of the patent issue. RAFI was originally involved in protesting against the worldwide collection of seeds by agricultural companies who selected for the most useful variants, obtained patent rights, and then attempted to sell them back to the countries of origin at a profit. RAFI saw the HGDP as an extension of these practices and rallied many of the indigenous populations in opposition to the HGDP.

Opposition from other scientists

Some scientists also believe the HGDP is poorly conceived. One vocal critic is Jonathan Marks, a Visiting Associate Professor of Anthropology at The University of California at Berkeley (CA, USA) who holds graduate degrees in both molecular genetics and anthropology. According to Marks, there are no 'pristine', genetically untouched, populations in the world. In his view, we are all mutts, and he doubts that the approach proposed by the HGDP will be as useful for studies of human evolution as proposed. He is concerned that the HGDP will encourage racism, and points out the difficulty of maintaining the confidentiality of the individuals being sampled. 'After all,' he says, 'if you have DNA then you have the ultimate identifier.' Marks believes that a study of genetic diversity would be a good thing, but is opposed to the methods of the HGDP. In his view, we should collect DNA along with extensive phenotypic data, even photographs, but only from individuals who have given informed consent and fully understand the implications of donating their DNA to the project. Marks believes that this could only be possible after a great deal of anthropological groundwork when collecting from indigenous populations, who are likely to have different ideas about blood, human bodies and heredity.

Should the HGDP be funded?

These issues recently came to a head when the National Research Council of the US National Academy of Sciences (NRC), which was asked to evaluate the proposed HGDP by the US National Science Foundation (NSF) and the NIH published its recommendations[9]. Their report is the culmination of four meetings, attended by a committee of 15 experts formed by the Board of Biology of the National Research Council. Three of the meetings were open to the public, and the committee heard from proponents of the project as well as from representatives of RAFI and indigenous peoples who opposed the project.

The report concluded that 'The federal government should provide funding for a global survey on human genetic diversity.' However, a statement from the committee that accompanied the report said 'The proposal does not clearly explain the purpose of the project or provide the necessary safeguards for protecting participants.' Because of this lack of focus of the existing proposal for the HGDP, the committee 'chose to examine the scientific merits and value of research on human genetic variation and the organizational, policy, and ethical issues that such research poses in a more-general context.' The remainder of the report is devoted to specific recommendations for conducting HGDP (summarized in Box 1), some of which are similar to those originally envisioned by the organizers of the project, while others, such as the strong emphasis on confidentiality, the ethics of informed consent and the arrangements regarding possible financial return, go further than the original HGDP proposal.

Box 1. Recommendations of the NRC's Commission on Life Sciences Regarding the Human Genome Diversity Project

1. Sampling strategy should record only the geographic location of the individual being sampled and the self-reported ethnicity, primary language, sex, age and parental birthplaces. The committee recommended against recording information on phenotypes (disease or otherwise), family pedigree, or any other information that would allow the sample to be linked to specific persons.
2. Blood samples collected from human populations should be converted into purified DNA and made available for widespread testing. It recommended against the designation of a common core of markers for genotyping in all samples.
3. A special panel of experts on materials and data management should be convened to address unresolved questions regarding questions of access to materials, data management and the enforcement of ethical guidelines. Some of these outstanding questions were enumerated by the Commission.
4. The freedom (including the right not to participate in the study), privacy and welfare of the individuals involved must be protected. A complete research protocol must be developed and divulged before obtaining consent. Women should be excluded from the study in regions where women's rights to self-determination are not recognized and where women do not have the freedom or power to choose whether to participate.
5. Arrangements regarding financial interests in the products or outcomes of the research should be negotiated as part of the original project review and informed-consent process.
6. US funding agencies should initially focus on funding projects originating in the USA, and expand their support to the international scene only after the US activities are successfully launched.

Different interpretations of the NRC report

However, it would seem that the report's recommendations were open to interpretation. *Nature*[10] and RAFI took the report to be a rejection of the HGDP's proposal, but *Science*[11] saw the report as supporting the project commenting that it 'got a cautious nod of approval from the National Research Council.' William J. Schull (University of Texas Health Center, Houston, TX, USA), the Chairman of the NRC review committee, says that it was not the intention of the committee to reject the proposal. In his words, the 'committee did not endorse or reject the HGDP.' Instead, it looked at the issue in a broad context to determine if the project has scientific merit and found that it has, but certain safeguards need to be put in place. To make sure the safeguards are followed, the committee recommended that initial funding be from US agencies only, since that would give them the greatest control over how the HGDP projects will be conducted. This statement was reiterated in a letter to *Nature*[12].

In December 1997, meetings were held at NIH and NSF with principal investigators who had received money for pilot projects related to techniques needed for the HGDP to go forward. The purpose of these meetings was to decide on the next step now that the NRC report is in hand. According to Mary Clutter, Assistant Director for Biological Sciences (National Science Foundation), the techniques to conduct the project are now in hand and the investigators have designed proposals that will meet the recommendations put forth in the NRC report. Clutter believes that the only approach to the continued objection to the project by some populations is to mount an education program to deal with the criticism.

Francis Collins, Director of the National Human Genome Research Institute of the NIH, was recently the lead author of an article proposing a project to catalog human DNA sequence variation[13]. The proposed project bears a striking resemblance to the HGDP; the article even cites the NRC report that endorses the HGDP project[9], but nowhere is the terminology 'Human Genome Diversity Project' used. The appearance of this article is a further suggestion that the recent NRC report might have broken through the impasse in the USA on funding for studies into variation of the human genome.

A revamped HGDP: but at what cost to medicine?

There seems to be little doubt that the HGDP, or an HGDP-like project under a different name, will soon move forward in the USA with funding of formal projects from the NIH and the NSF. However, the NRC report has imposed restrictions that might decrease the value of the DNA samples for medical research. For example, it is difficult to understand how the samples could be used to study adult-onset diabetes in selected populations if it is not possible to record which samples come from individuals with diabetes. The lack of pedigree information and the requirement that informed consent must be given for defined protocols might also prove restrictive. The review committee must have recognized that such restrictions limit the medical value of the samples, but concluded that the risk of exploitation was sufficiently great to require a sacrifice of the medical value of the samples.

It remains to be seen how HGDP funding agencies in other countries will respond to the report. Sir Walter Bodmer notes that the NRC report 'will have nothing to do with what might, or might not, go on elsewhere.' It is also unlikely to have an affect on what goes on as biotechnology companies conduct their own miniature, self-funded versions of the HGDP. In the midst of all the conflict, they have quietly stepped in, and are quite likely to be the ones to reap the major medical benefits originally envisioned by the HGDP. ❑

Questions arising for molecular medicine
• Will the proposed HGDP receive funding in the USA now that the NRC has issued an overall favorable report?
• Will the constraints put on the project severely compromise the research value of the genetic material that is to be collected?
• What does informed consent mean to populations of people who have no knowledge of modern medicine or conceptual basis to understand the implications of the collection and study of their genetic material? If an educational effort is to be undertaken to deal with this question, who should fund and direct such an effort?
• How will the clash between the desire of scientists to collect genetic material and the view of some indigenous peoples regarding the sacredness of such material be resolved?
• Who will have access to the DNA collected under the HGDP, how and by whom will it be distributed, and how can all possible experimental protocols be anticipated, as recommended by the NRC report as part of the informed consent statement?
• Should biotechnology companies who collect DNA for development of commercial products be required to work under the same set of guidelines recommended for the HGDP?

References

1. Cavallli-Sforza, L.L. *et al.* (1991) **Call for a worldwide survey of human genetic diversity: a vanishing opportunity for the human genome project.** *Genomics* 11, 490-498
2. Greely, H.T. (1996) **Genes, patents, and indigenous peoples,** *Cultural Survival Quarterly* 20, 54-57
3. Knowler, W.C. *et al.* (1993) **Determinants of diabetes mellitus in Pima Indians,** *Diabetes Care* 16, 216-227
4. Risch, N. and Botstein, D. (1996) **A manic depressive history,** *Nat. Genet.* 12, 351-353
5. Lehrman, S. (1997) **Jewish leaders seek genetic guidelines...,** *Nature* 389, 322
6. Fowke, K.R. *et al.* (1996) **Resistance to HIV-1 infection among persistently seronegative prostitutes in Nairobi, Kenya,** *Lancet* 348, 1347-1351
7. Marshall, E. (1997) **Tapping Iceland's DNA,** *Science* 278, 566
8. Maybury-Davis, D. (1996) **Science and sensibility,** *Cultural Survival Quarterly* 20, 3
9. Committee on Human Genome Diversity, Commission on Life Sciences, National Research Council (1997) *Evaluating Human Genetic Diversity*, National Academy Press, Washington, DC
10. Macilwain, C. (1997) **Diversity project 'does not merit federal funding',** *Nature* 389, 774
11. Pennisi, E. (1997) **NRC Oks long-delayed survey of human genome diversity,** *Science* 278, 568
12. Schull, W.J. (1997) **Support for Genetic Diversity Project,** *Nature* 390, 221
13. Collins, F.S., Guyer, M.S. and Chakravarti, A. (1997) **Variations on a theme: cataloging human DNA sequence variation,** *Science* 278, 1580-1581

* * * * * * * * * * * * * * * * * * * *

Questions:

1. What were the ethical guidelines agreed upon by the HGDP investigators?

2. Name four populations of people where adult-onset diabetes is strongly familial.

3. What was the reaction from many of the populations of the world when they heard about the prospects of having their DNA collected and studies?

Answers are at the end of the book.

- 28 -

Ever since the genetic code was cracked by Watson and Crick in 1953, there has been considerable debate as to its origin. Some biologists argue that the key to translating DNA into proteins is merely an accident that has been resilient to the test of time, while others maintain that natural selection plays a role in the code. Researchers at the annual meeting of the Society for the Study of Evolution report that factors other than change helped shape the code's origin and history. The researchers hypothesized, and now have data to suggest, that before cells had the equipment to translate the code and build proteins, it could have been shaped by affinities between specific base sequences and amino acids. These experiments coupled with computer analysis provide evidence that the genetic code resists errors better than a million other possible codes. Could it really be an accident?

Tracking the History of the Genetic Code

by Gretchen Vogel

Science, July 17, 1998

Computer analyses and experiments with RNA molecules offer new insight into the forces that may have shaped the genetic code over time.

Vancouver — For the 3 decades since biologists cracked the genetic code — the key to translating DNA into proteins — they have debated its origins. Some claimed it must be a random accident forever frozen in time, while others argued that the code, like all other features of organisms, was shaped by natural selection. Most of those debates have been philosophical, with little data to back up either side. But at the annual meeting of the Society for the Study of Evolution held here last month, two speakers presented evidence suggesting that forces other than chance shaped the code's origin and history.

Experiments with RNA have shown that chemical attractions between the genetic material and the components of proteins may have helped shape the original code, reported one speaker. Another researcher, using powerful computer analyses, suggested that the modern code is the product of evolution because it is so error-proof: Only one in a million other possible codes is better at producing a workable protein even when the DNA carries mistakes.

Doubters such as evolutionary biologist Niles Lehman of the State University of New York, Albany, still remain unconvinced that the code is anything but an accident. But he and others say that new studies such as these, as well as other work probing the history of individual genetic "words" are beginning to make a dent in their skepticism. "We're at a turning point" for probing the origins and history of the code, says Lehman.

Living things use DNA to store the instructions for making the proteins that build cells and direct them to develop into a complete organism. The four different subunits, or bases, that make up the DNA chain are grouped into three-letter "words" called codons, and each codon specifies a protein's amino acid building block. Specialized cellular machinery copies the DNA code into RNA — which has a similar code — and then reads the RNA to piece together the amino acids to make proteins. A codon "means" the same thing in a koala as it does in a rose or a bacterium. Yet there's no clear pattern in the pairing of codons and amino acids, which has persuaded many scientists that the code arose by accident.

But test tube experiments now suggest that before cellular machinery had evolved to read the code and build proteins, the code could have been shaped by affinities between specific base sequences and amino acids. Many scientists have speculated about such a scenario, but new data from experiments in which short strands of RNA are chosen based on their affinity for an amino acid are allowing them to test the idea. Several years ago, Michael Yarus of the University of Colorado, Boulder, noticed that in his experiments, the RNA strands that were best at binding a given amino acid tended to contain codons for that amino acid. But because the three-base codons often show up at random, the data were inconclusive.

Now evolutionary biologist Laura Landweber and graduate student Rob Knight of Princeton University have done a more careful analysis, looking specifically at where the amino acid arginine binds to random RNA strands generated in several researchers' experiments. If there is no real affinity, they reasoned, codons for arginine will appear as often in the regions where the amino acid does not bind as in regions that arginine homes in on. They found, instead, that while arginine codons made up 30% of the nonbinding RNA sites — the expected percentage, given that arginine has many possible codons — they made up 72% of the sequenced in the binding regions. That suggests, says Landweber, that it's no accident that these codons specify arginine.

The arginine evidence is intriguing, says evolutionary biologist Leslie Orgel of the Salk Institute in La Jolla, California. "But it's premature to draw any very strong conclusions" from data on the affinities of a single amino acid,

he says. Researchers are delighted, however, that experimenters are now tackling the question. "Previously we had to rely solely on theory," says Lehman, but "if [Landweber's analysis] holds up, it will provide a convincing body of evidence" that basic chemical forces helped to shape the code.

Once the code was born, a different kind of pressure, the need to minimize errors, might have refined it. While some researchers have argued that any changes to the code over its 3.5-billion-year history would have been like switching the keys on a typewriter, leading to hopelessly garbled proteins, others argued that the existing code is so good at its job that it must have been shaped by natural selection. For example, in 1991, evolutionary biologists Laurance Hurst of the University of Bath in England and David Haig of Harvard University showed that of all the possible codes made from the four bases and the 20 amino acids, the natural code is among the best at minimizing the effect of mutations. They found that single-base changes in a codon are likely to substitute a chemically similar amino acid and therefore make only minimal changes to the final protein.

Now Hurst's graduate student Stephen Freeland at Cambridge University in England has taken the analysis a step farther by taking into account the kinds of mistakes that are most likely to occur. First, the bases fall into two size classes, and mutations that swap bases of similar size are more common than mutations that switch base sizes. Second, during protein synthesis the first and third members of a codon are much more likely to be misread than the second one. When those mistake frequencies are factored in, the natural code looks even better: Only one of a million randomly generated codes was more error-proof.

That suggests, Freeland says, that the code has been optimized over the eons and isn't simply the product of chance. Lehman agrees that the one-in-a-million result looks impressive, but cautions that the statistics could be misleading. A high degree of similarity within one clan of amino acids could account for the code's apparent resistance to error, and the rest of the code could be random, he says.

With both the genesis and history of the code

looking less and less accidental, Landweber and Freeland plan to collaborate new year, hoping to "build a grand scheme of the code's raison d'être," Landweber says — whether it be accident or design. ❑

* * * * * * * * * * * * * * * * * * * *

Questions:

1. Does DNA get translated into RNA or is RNA translated into DNA?

2. Which amino acid was the key to understanding the affinity model?

3. What were Freeland's findings when he investigated the single-base change experiments further?

Answers are at the end of the book.

- 29 -

Traditionally, the terms 'Darwinism' and 'Lamarckism' mean very different things to different people. Lamarck's theory proposed that simple life forms have the ability to spontaneously regenerate themselves. His theory also stated progressive change was caused by the intrinsic tendency of living organisms to become more complex and that mutation results by the increased use and disuse of parts in response to environmental conditions, and these modifications were inherited. This contrasted to Darwin's theory that the mechanism leading to evolutionary change was natural selection of heritable variations, and Darwin went on to prove his theory by explaining the origin and behavior of inherited variations. He agreed with Lamarck's idea that when traits change through use or disuse those traits tend to be inherited, but he believed that new environments increase the rate at which variation is generated. This article discusses multiple inheritance systems all with properties that follow Darwin's evolutionary theory.

'Lamarckian' Mechanisms in Darwinian Evolution

by Eva Jablonka, Marion J. Lamb, and Eytan Avital

Trends in Ecology & Evolution, May 1998

Since the Modern Synthesis, evolutionary biologists have assumed that the genetic system is the sole provider of heritable variation, and that the generation of heritable variation is largely independent of environmental changes. However, adaptive mutation, epigenetic inheritance, behavioural inheritance through social learning, and language-based information transmission have properties that allow the inheritance of induced or learnt characters. The role of induced heritable variation in evolution therefore needs to be reconsidered, and the evolution of the systems that produce induced variation needs to be studied.

'Lamarckism' and 'darwinism' are traditionally seen as alternative theories trying to account for evolutionary change (Box 1). The verdict of history is that Lamarck got it wrong — evolutionary change does not occur through the inheritance of acquired characters. Acquired characters are the outcome of *instructive processes,* such as those seen in embryonic induction, transcriptional regulation, and learning, all of which involve highly specific and usually adaptive responses to factors external to the responding system. The inheritance of the outcomes of instructive processes is deemed to be impossible. Adaptive evolutionary change is assumed to be based on darwinian (or more accurately neo-darwinian) evolution in which guidance comes exclusively from selective processes. The production and nature of heritable variation is assumed to be uninformed by the environment or by previous history. The future is open-ended, determined solely by the contingencies of life. It is neither foretold nor intimated.

General selection theory makes no assumptions about the origin of heritable variation. It maintains that evolution by natural selection will occur in any system with entities manifesting the properties of multiplication, heredity and heritable variation affecting reproductive success[1]. In the current version of biological darwinism, it is assumed that information is digital and encoded in DNA base sequences, that multiplication of information occurs through DNA replication, and that variation, which is generated by mutation and recombination, is random with respect to the selecting environment and the developmental history of the organism and the lineage. However, this version of evolution — 'genic neodarwinism' — is incomplete: it gives natural selection an exclusive deterministic role in the evolution of all conceivable complex adaptations, but until recently it has had rather little to say about the evolution of new systems for acquiring, storing and transferring information, and even less about the evolutionary effects of such systems once they are in place. Natural selection leads not only to the evolution of eyes, wings, and sonars, but also to the evolution of new evolutionary rules. Many of these rules undermine the assumption that variation is random. Mechanisms allowing the inheritance of acquired characters have evolved several times during the history of life, and understanding their evolution is crucial to understanding the transitions to new levels of individuality[2-4].

Evolved 'lamarckian' heredity systems

The heredity systems that we consider here are all complex mechanisms for the acquisition, storage and transfer of information. All evolved through natural selection, but they differ from each other in the type of information they transmit, in their evolutionary history, and in their evolutionary effects. They include adaptive mutational systems involving non-random changes in DNA, cellular heredity systems in which information is acquired and transmitted through intracellular structures and biochemical mechanisms, the transfer of patterns of behaviour through social learning coupled with certain types of social organization, and the transmission of information using symbolic languages. All of these systems allow certain outcomes of the interaction between the organism and its environment to be incorporated into and maintained within the information-carrying system, and the information to be transmitted to future generations. All therefore allow the inheritance of acquired or learnt characters.

Adaptive mutational systems: the intelligent genome

In the genetic inheritance system, information for making RNA and proteins is stored in DNA base sequences; an elaborate enzyme system enables this information to be replicated and transmitted to the next generation. Physico-chemical damage to the DNA and errors occurring during its replication can be removed by a battery of repair processes. Errors that remain, and sequence changes that are created during repair, or result from the movements of genomic parasites, provide the raw material on which selection ultimately acts. Since replication and repair are enzyme-dependent processes, genetic variation in the enzyme-coding genes can affect their efficiency. In addition, base sequences differ in the likelihood that they will be damaged, replicated inaccurately, or be invaded by parasitic elements. Consequently, the type of DNA variation generated, the rate at which it is generated, and when and where it is generated, can all be selectively modified. Systems that were presumably selected to maintain the fidelity of the genetic inheritance system can, once they are in place, be modified and used in ways that lead to the generation of mutations at particular sites, or of a particular type, or at a particular time.

Some pathogenic bacteria are able to adapt to ever-changing conditions because genes that are likely to provide the means of surviving are hypermutable. Moxon and his colleagues have shown that the pathogen *Haemophilus influenzae* can adapt to the changing conditions experienced in its host because genes influencing its antigenicity and other features are highly mutable[5]. These genes are characterized by tandem repeats of short DNA sequences which make them prone to genetic change through homologous recombination and strand slippage. Through selection, the genome of *H. influenzae* has evolved features that result in high rates of variation in the relevant genes.

Similar highly targeted, locus-specific DNA changes are found in trypanosomes and other microbial pathogens.

Bacteria are able to generate variation not only at sites, but also at times when it is likely to be selectively useful. During starvation, some mutations arise at a higher rate than in non-starving cells. This could simply be because starved cells cannot carry out normal repair. However, it is increasingly clear that bacteria have evolved mechanisms enabling them to actively generate new variation in conditions where survival depends on genetic change. Shapiro found that one type of mutation in *Escherichia coli, araB-lacZ* fusions, involves a multi-step process that is physiologically regulated[6]. Recently, Wright has shown that amino acid starvation of *E. coli* increases transcription of a specific set of genes that enable them to survive, and these genes have enhanced mutation rates[7]. Genetic changes induced by altered environments have also been found in cultured plant and animal cells, and germline-transmissible changes have been well characterized in flax[8] and mustard[9].

Shapiro suggests that organisms respond to stress by activating their 'natural genetic engineering systems'[6]. There is biological and environmental feedback into the genome. This means that evolutionary change may be very rapid because mutation rates can be increased, and coordinated changes may occur at many sites within a single genome. Furthermore, although the induced genetic changes may not be specifically those that solve the organism's immediate survival problem, if similar stress episodes have been frequent in the past, the genome and genetic machinery may have been modified to target variation to a subset of sites or to be of a type that is likely to provide useful variation. Whether or not mutations is 'directed' in the sense that environmental factors induce exclusively those mutations that are beneficial is still highly debatable. However, given the well-known ability of natural selection to take two or more old functions and cobble them together to construct something new, it would not be surprising if the mechanisms that enable selective control of transcription have been coupled with those that maintain (or fail to maintain) genetic fidelity. The result would be that the inducible systems that turn genes on and off could then also turn the production of genetic change on and off.

We know that in some organisms the genetic system has been modified in ways that make the genome change in a directed way: in ciliates, formation of the macro-nucleus involves a complex, regulated series of sequence deletions, duplications and rearrangements; in amphibian oogenesis, rRNA genes are selectively amplified; in *Ascaris* development, large parts of the genome are deleted in somatic cells; in *Drosophila*, chorion genes are amplified in follicle cells; in the mammalian immune system, complex developmentally regulated DNA rearrangements followed by hyper-mutation underlie the production of antibody diversity[10]. The programmed generation of variation seen in these organisms shows that the genetic inheritance system can take part in instructive processes.

Cellular heredity: epigenetic inheritance systems

From what we know at present, it seems that only rarely is DNA sequence information changed during determination and differentiation. Most organisms have different ways of generating and transmitting cellular phenotypes — they use epigenetic inheritance systems (EISs). Three chasses of EIS have been recognized: steady-state, structural, and chromatin-marking systems[11].

Steady-state systems are based on positive feedback loops. At its simplest, a gene produces a product that stimulates further activity of the gene and hence further synthesis of the product. Once switched on by physiological or developmental events, the cell lineage continues transcription unless the concentration of the product falls. In structural inheritance systems, cell structures are used to template the formation of new similar structures. For example, genetically identical ciliates can have different patterns of cilia on their cell surface which are inherited; even experimentally altered patterns can sometimes be transmitted to daughter cells[12]. Recent studies of prions provide further examples of structural inheritance[13]. In the third type of EIS, chromatin-marking, states of chromatin that affect gene expression are clonally inherited. The textbook example is the transmission of the inactive X chromosome in female mammals; once one of the two X chromosomes in a cell has been inactivated, all of the descendants of that cell normally have the same X inactive. Information about gene and chromosome activity is contained in what have been called chromatin marks, i.e., in the proteins associated with DNA and the distribution of DNA modifications such as cytosine methylation[14].

We can only guess at the evolutionary origins of the various EISs. Simple feedback and structural inheritance systems probably evolved very early in evolutionary history, but chromatin-marking systems depend upon DNA and its orderly replication, so they must have followed the evolution of DNA-based inheritance. DNA methylation and heterochromatinization are both ways of inactivating repeated DNA sequences such as those that arise from viral invasion, so the chromatin-marking EISs may have originated as defense mechanisms against genomic parasites which were later modified by selection to fulfil additional roles in gene regulation and cell memory[15,16].

Whatever the origins of the various EISs, they were probably selectively refined in early unicells living in environments that fluctuated in a regular way (e.g., between summer and winter)[11,17]. When environmental fluctuations are short relative to the lifespan, selection favours adjustment through physiological change (for example, turning genes on and off); when they are long, so that many generations are spent in each phase of the cycle, adaptation through the selection of genetic variation is possible. However, in cycles of intermediate length, where adaptation through selection of genetic variations is usually too slow (although hypermutable loci may evolve), the ability to transmit functional states (for example, whether a gene is on or off) should be strongly selected. Any genetic changes that link epigenetic switching to the environmental change would have obvious selective advantages.

The outcome of this type of selection was

systems through which alternative cellular phenotypes, including induced phenotypes, could be transmitted in cell lineages with various degrees of fidelity. Such systems opened up new evolutionary possibilities. EISs were probably important in the transition to multicellularity[2,18], because the phenotypic uniformity they gave to a clonal group of cells made the variation within groups less than variation between groups. Certainly, complex ontogenies would have been unlikely without EISs, since the ability of cells to transmit their induced epigenetic states to daughter cells is fundamental to development. In organisms that reproduce by budding or fragmentation, induced changes in epigenetic states can readily be transmitted to further generations. In such organisms inherited epigenetic variations may be able to 'hold' an adapted state for long enough to allow similar genetic variations to catch up. Theoretically, transmitting induced phenotypes, even if only for a few generations, can have considerable advantages in some environments[19].

Social learning and the origin of traditions

Animals acquire information that affects the way they will behave in the future and store it in their nervous systems. In animals that show parental care and other forms of social interaction, patterns of behaviour can be transferred between individuals and across generations. New patterns of behaviour, first acquired either by accident or by individual learning in new conditions, can be transmitted transgenerationally through social learning[20]. This inheritance system operates at the whole organism level, and the information encoded is analog in nature — as a rule, it is not readily dissociated into independently heritable parts, but is contained in the dynamics of the interactions between the organism and its social and ecological environment. It is this dynamic system that is reconstituted anew every generation.

In birds and mammals, social learning occurs when the presence of one relatively experienced individual increases the chances that a naive individual learns a similar behaviour pattern. A young male song-bird, hearing the song of his father, learns some of the idiosyncratic components of the song, and will later transmit it to his sons[21]. Naive blackbirds, seeing the mobbing behaviour of conspecifics towards a particular target, learn to mob this type of object, even when it is harmless[22]. One of the best-studied cases of social learning in mammals is Aisner's and Terkel's study of cultural inheritance of a novel feeding habit in Israeli black rats (*Rattus rattus*)[22,24]. These rats have recently extended their range to include the Jerusalem-pine forest, where the only source of food is the pine seeds enclosed within inedible pine cones. The rats reach the nutritious seeds by an elaborate cone-stripping technique. Results of cross-fostering experiments, where pups of stripping mothers were raised by non-strippers and vice versa, have shown that the ability to strip pine cones efficiently is learnt and not genetically determined.

Social learning can have important effects on the evolution of behaviour, and hence on our interpretation of evolutionary change. First, when we observe a new heritable pattern of behaviour in a population, its origin and maintenance cannot automatically be assumed to be due to genetic variations. The inheritance of purely 'cultural variations' has to be seriously considered. Second, taking behavioural inheritance into consideration leads to alternative or complementary interpretations of known patterns of behaviour. For example, it is possible to show that adoption can spread and be maintained within populations through the social learning of certain parenting styles[25]. Third, the possibility that sexual imprinting and other mechanisms of social learning can initiate speciation must be considered[26].

New acquired habits and traditions, such as pine cone-stripping, often precede genetic adaptations, and exert persistent directional selection for genetic variations that are congruent with the new pattern of behaviour. For example, the continual use of pine seeds as a major source of food gives a selective advantage to any genetic variation that improves the finding, processing, and digestion of pine seeds. In addition, selection for the ability to learn this particular behaviour pattern ever more efficiently and rapidly may eventually lead to it becoming dependent on fewer learning trials, or even on no learning trials at all — in short, it may lead to partial or full genetic assimilation (Box 2)[27].

Symbolic languages

Language is part of the behavioural inheritance system. However, some properties of language make it qualitatively different from any other information storage and transmission system: language has a structure, or syntax, that allows generativity and creativity. Utterances are organized into sentences by applying rules for the formation of hierarchical, recurring constructions of meaning-relations. This leads to the production and comprehension of an infinite number of meaningful sentences. Moreover, language increases the scope of transmissible information: the information is not only messages about the world, but also messages about beliefs, about past and future, and about things distant, abstract, and absent. Unlike other types of information transmitted through the nervous system, linguistic information is organized digitally, and is decomposable into semantic units. The differences between individuals in the ability to use a particular language, and the differences between languages, show how plastic the system is, notwithstanding the fact that language has an evolved genetic basis[28].

Reconstructing the evolution of language is extremely problematical, not only because we have to rely on very indirect evidence, but also because there is disagreement among linguists as to what qualifies as 'true' language. The ape-language controversy illustrates the problem: Kanzi, the language-instructed bonobo (*Pan paniscus*), can (arguably) understand some form of simple spoken English, and is able to converse, though with little syntactical regularity and with the aid of human-made visual symbols, about his desires and intentions[29]. However, many linguists are adamant that the non-syntactical communication and representation system of Kanzi and other language-instructed higher apes are not a true language, but rather a 'proto-language'[30]. The acquisition of syntax, which is unique to humans, is perceived by many linguists as a saltatory event. However, to most evolutionary biologists it seems more likely that the phonetic system, the ability to learn many lexical elements, and syntax must all have evolved through gradual cultural and genetical evolution. Eventually, from some form of proto-language, the mature language of *Homo sapiens* evolved.

Certain basic preconditions for language evolution were presumably present in our ancestors, since they exist in the higher apes, and also in some monkeys. These include intense sociality with a highly evolved social intelligence involving some theory of mind (the ability to understand another individual's intentions), a sign system that functions to modulate social interactions, and increased voluntary motor control of hands, breathing, and the expression of emotions[31]. Dunbar has stressed the importance of social bonding among individuals for early proto-language evolution[32]. He argued that as human groups grew in size, grooming was no longer able to manage the intimate and complex social interactions within the group. Language evolved as a more effective grooming-system, because it involves several individuals at the same time. Other authors have stressed other features that have been important in the evolution of language, such as its prosodic and rhythmical aspects which lead to group-bonding[33], and its role in extending and allowing voluntary access to memories[34].

The reconstruction of the evolution of language involves reconstructing the evolution of a 'linguistic community', for language is a community activity. The gelada baboons (*Theropithecus gelada*), whose rhythmical and melodious vocalizations accompany every act of their busy social lives[35], might be a good model for such a community. If we add to the gelada 'vocal' community the more sophisticated intellectual capacity of higher apes, and the use of vocalization for the transfer of information about specific events, such as vervet monkeys' (*Cercopithecus aethiops*) predator-specific alarm calls[36], we have a preliminary linguistic community. The linguistic abilities of Kanzi give us some idea as to how much could have been achieved by cultural evolution without genetic changes. The transition from proto-language to a full syntactical system may be especially difficult to reconstruct, but

contrary to the opinions of those linguists who envisage the sudden emergence of mature syntactical language from proto-language[37], it is possible to construct a plausible detailed scenario of this transformation, as recently shown by Aitchison[38].

The evolution of language may have been driven by cultural evolution, followed by genetic assimilation. Genes and culture co-evolved, with the rare linguistic innovations of adults being quickly adopted and absorbed by the young children of the group, to become the behavioural norm and habit of the next generation[39]. Cultural evolution for ever more efficient linguistic competence 'stretched' the linguistic abilities of the individuals and exerted stable and consistent directional selection on the underlying, genetically-encoded features of the nervous system that promote every more effective language use[40].

Conclusions

All the inheritance systems that we have described have the properties necessary for darwinian evolution: the information (be it digital or non-digital) can vary, and the variations can be transmitted across generations. Natural selection of the variations leads to adaptation and divergence. Although variations may result from random errors in maintenance and copying processes, heritable variations in these systems (1) are often induced or learnt, (2) are frequently adaptive, (3) can sometimes be generated at a very high rate and are often reversible, (4) may affect several characters in the same individual, and (5) may affect several individuals in a similar way. Rapid, reversible, co-ordinated, induced or learnt changes are not characteristics of the classical genetic system. Although the evolved systems allowing 'adaptive' mutation obviously depend on DNA variations, DNA base sequence changes are not required for the generation of epigenetic, behavioural, and language-based variations.

It is possible to study the effects of each inheritance system in isolation from the others for a limited number of generations, but the different systems clearly interact. One of the most neglected and challenging aspects of the study of

139

evolutionary history is the investigation of these interactions. There is a simple schematic illustration of the interrelationships. Genetic adaptations may be guided by heritable induced or learnt phenotypic adaptations, usually through genetic assimilation. As additional inheritance systems evolved, the new systems began to guide the changes at the underlying levels. The importance of the different systems is not the same in all groups. While EISs and genomic mutational systems have probably played a major role in the evolution of unicellular organisms, plants, fungi and lower animals, behavioural inheritance systems have attained prominence in higher animals, and language has a major directing role in human evolution.

Instructive (or 'lamarckian') inheritance systems are all adaptations that have evolved through darwinian natural selection, often through the selection of randomly generated variations. The kind of environment in which such systems have evolved is likely to have been one with some re-occurring features, but with enough temporal or spatial diversity to preclude a fixed genetic response on the one hand, yet on the other hand make an individual, physiological response too costly[16,17,41]. Once present, these instructive inheritance systems constrain and channel evolution. Unlike other constraints on evolutionary change, such constraints do not merely define the range of the possible, they also, more positively, specify what is likely. The future may spring more directly from the present than we have been accustomed to believe. ❑

Box 1. Lamarckism and Darwinism

The terms 'darwinism' and lamarckism' mean different things to different people[16]. Nowadays, they are rarely used in a historically correct way. Lamarck's theory of evolution was wide-ranging and included continual spontaneous generation of simple forms of life, progressive change caused by the inherent tendency of living matter to become more complex, modifications brought about by the increased use and disuse of parts in response to environmental conditions, and the inheritance of such modifications. In Darwin's theory the key mechanism leading to evolutionary change was natural selection of heritable variations, and Darwin put forward his pangenesis theory to explain the origin and behaviour of inherited variation. He accepted that when characters change through use or disuse the changes are inherited, and argued that new environments increase the rate at which variation is generated.

By the end of the 19th century, thanks largely to the efforts of Weismann, neodarwinians had purged darwinian evolution of pangenesis and the inheritance of acquired characters. Following the synthesis of genetics and darwinian ideas in the 1930s and 1940s, darwinism came to mean the theory that evolutionary change results from natural selection of the random genetic variations generated by mutation and recombination. The neolamarckians of the late 19th and early 20th centuries emphasized various parts of Lamarck's theory, but in modern biology lamarckism is usually equated with the inheritance of acquired characters. Other parts of the Darwin's and Lamarck's theories are usually ignored. In this article we use the terms darwinism and lamarckism in the familiar colloquial sense, rather than in a historically precise way.

Box 2. Genetic assimilation through the culturally-driven Baldwin effect

The transformation, through darwinian selection, of a learnt response into a fixed or 'instinctive' response is known as the 'Baldwin effect', although two other late nineteenth-century scientists, Lloyd Morgan and Fairfield Osborne, suggested a similar mechanism. The basic idea is clearly expressed by Morgan: '...any hereditary variations which coincide in direction with modifications of behaviour due to acquired habit would be favoured and fostered: while such variations as occurred on other and divergent lines would tend to be weeded out. ... We may look upon some habits as the acquired modifications which foster those variations which are coincident in direction, and which go to the making of instinct.' (cited by Hardy[27], p. 197).

The British embryologist and geneticist, C.H. Waddington, suggested a similar process, genetic assimilation, for the evolutionary acquisition of some physiological and developmental adaptions. Using *Drosophila*, Waddington showed that a character whose development originally depended on an environmental stimulus became, through selection in the inducing environment, genetically fixed and began to appear in the normal non-inducing environment. This happened through the sexual reshuffling of genes and selection of those combinations that produced progressively more rapid and efficient responses to the stimulus[42].

Social learning may augment the probability of genetic assimilation of behaviour patterns. Tradition makes selection more directional and more persistent because it leads to the transmission of the same selective regime. A behaviour pattern that becomes traditional through social learning is often more enduring than that acquired by individual learning because it results from the self-perpetuating social structure, and may continue even when the environmental conditions that first initiated it in a particular individual have changed.

References

1. Maynard Smith, J. (1986) *The Problems of Biology*, Oxford University Press
2. Jablonka, E. (1994) **Inheritance systems and the evolution of new levels of individuality.** *J. Theor. Biol.* 170, 301-309
3. Jablonka, E. and Szathmáry, E. (1995) **The evolution of information storage and heredity,** *Trends Ecol. Evol.* 10, 206-211
4. Maynard Smith, J. and Szathmáry, E. (1995) *The Major Transitions in Evolution*, Freeman
5. Moxon, E.R. *et al.* (1994) **Adaptive evolution of highly mutable loci in pathogenic bacteria,** *Curr Biol.* 4, 24-33
6. Shapiro, J.A. (1997) **Genome organization, natural genetic engineering and adaptive mutation,** *Trends Genet.* 13, 98-104
7. Wright, B.E. (1997) **Does selective gene activation direct evolution?** *FEBS Lett.* 402, 4-8
8. Cullis, C.A. (1984) **Environmentally induced DNA changes,** in *Evolutionary Theory: Paths into the Future* (Pollard, J.W., ed), pp. 203-216, Wiley
9. Waters, E.R and Schaal, B.A. (1996) **Heat shock induces a loss of rRNA-encoding DNA repeats in *Brassica nigra*,** *Proc. Natl. Acad. Sci. U.S.A.* 93, 1449-1452
10. Alberts, B. *et al.* (1989) *Molecular Biology of the Cell (2nd edn)*, Garland
11. Jablonka, E., Lachmann, M. and Lamb, J.J. (1992) **Evidence, mechanisms and models for the inheritance of acquired characters,** *J. Theor. Biol.* 158, 245-268
12. Grimes, G.W. and Aufderheide, K.J. (1991) **Cellular aspects of pattern formation: the problem of assembly,** *Monogr. Dev. Biol 22*, Karger
13. Tuite, M.F. and Lindquist, S.L. (1996) **Maintenance and inheritance of yeast prions,** *Trends Genet.* 12, 467-471
14. Jablonka, E. and Lamb, M.J. (1989) **The inheritance of acquired epigenetic variations,** *J. Theor. Biol.* 139, 69-83
15. Bestor, T.H. (1990) **DNA methylation: evolution of a bacterial immune function into a regular of gene expression and genome structure in higher eukaryotes,** *Philos. Trans. R. Soc. London Ser. B 326*, 179-187
16. Jablonka, E. and Lamb, M.J. (1995) *Epigenetic Inheritance and Evolution: the Lamarckian Dimension*, Oxford University Press
17. Lachmann, M. and Jablonka, E. (1996) **The inheritance of phenotypes: an adaptation to fluctuating environments,** *J. Theor. Biol.* 181, 1-9.
18. Jablonka, E. and Lamb M.J. **Epigenetic inheritance in evolution,** *J. Evol. Biol.* (in press)
19. Jablonka, E. *et al.* (1995) **The adaptive advantage of phenotypic memory in changing environments,** *Philos. Trans. R. Soc. London Ser. B* 350, 133-141
20. Heyes, C.M. and Galef, B.G. (eds) (1996) *Social*

Learning in Animals: the Roots of Culture, Academic Press

21. Kroodsma, D.E. and Miller, E.H. (eds) (1996) *Ecology and Evolution of Acoustic Communication in Birds,* Cornell University Press

22. Curio, E., Ernst, U. and Veith, W. (1978) **Cultural transmission of enemy recognition: one function of mobbing,** *Science* 202, 899-901

23. Aisner, R. and Terkel, J. (1992) **Ontogeny of pine cone opening behaviour in the black rat,** *Rattus rattus, Anim. Behav.* 44, 327-336

24. Terkel, J. (1996) **Cultural transmission of feeding behavior in the black rat (*Rattus rattus*),** in *Social Learning in Animals: the Roots of Culture* (Heyes, C.M. and Galef, B.G., eds), pp. 17-47, Academic Press

25. Avital, E., Jablonka, E. and Lachmann, M. **Adopting adoption,** *Anim. Behav.* (in press)

26. Laland, K.N. (1994) **On the evolutionary consequences of sexual imprinting,** *Evolution* 48, 477-489

27. Hardy, A. (1965) *The Living Stream,* Collins

28. Jackendoff, R. (1993) *Patterns in the Mind,* Harvester Wheatsheaf

29. Savage-Rumbaugh, S. and Lewin, R. (1994) *Kanzi: The Ape at the Brink of the Human Mind,* Doubleday

30. Bickerton, D. (1990) *Language and Species,* University of Chicago Press

31. Gibson, K.R. and Ingold, T. (eds) (1993) *Tools, Language and Cognition in Human Evolution,* Cambridge University Press

32. Dunbar, R. (1996) *Grooming, Gossip and the Evolution of Language,* Faber & Faber

33. Ujhelyi, M. (1996) **Territorial song and long call: language precursors?** in *Origins of Language* (Trabant, J., ed.), pp. 142-154, Collegium Budapest Workshop Series No. 2, Collegium Budapest

34. Donald, M. (1991) *Origins of the Modern Mind: Three Stages in the Evolution of Culture and Cognition,* Harvard University Press

35. Richman, B. (1987) **Rhythm and melody in Gelada vocal exchanges,** *Primates* 28, 199-223

36. Cheney, D.L. and Seyfarth, R.M. (1990) *How Monkeys See the World,* University of Chicago Press

37. Chomsky, N. (1965) *Aspects of the Theory of Syntax,* MIT Press

38. Aitchison, J. (1996) *The Seeds of Speech: Language Origin and Evolution,* Cambridge University Press

39. Morgan, E. (1994) *The Descent of the Child: Human Evolution from a New Perspective,* Souvenir Press

40. Jablonka, E. and Rechav, G. (1996) **The evolution of language in light of the evolution of literacy,** in *Origins of Language* (Trabant, J., ed.) pp. 70-88, Collegium Budapest Workshop Series No. 2, Collegium Budapest

41. Boyd, R. and Richerson, P.J. (1985) *Culture and the Evolutionary Process,* University of Chicago Press

42. Waddington, C.H. (1957) *The Strategy of the Genes,* Allen & Unwin

* * * * * * * * * * * * * * * * * * * *

Questions:

1. How is the current version of biological Darwinism incomplete?

2. How are some pathogenic bacteria genetically altered to withstand their constantly changing environment?

3. What are the three classes of epigenetic inheritance systems?

4. True or False? Syntax in language is not unique to humans.

Answers are at the end of the book.

Part Eight

Behavioral Genetics

30

There are many research studies taking place each year that focus on the identification of novel genes for the many unexplained inherited conditions. Many of the families that are affected with these conditions are eager to help scientists discover the cause of the condition that plagues their family and are highly motivated to help discover a cure. However, sometimes during the pursuit of an answer to these perplexing conditions, the confidentiality and privacy of the research participants is forgotten. Jeffrey Botkin, from the university of Utah, completed a study investigating the approach that investigators and journals use to guarantee participant confidentiality and privacy. His study suggests that standard policies need to be implemented to help protect families involved in research but still allow for scientific publications and advancement.

Concerns Mount over Privacy As Genetic Research Advances

by Stephen P. Hoffert

The Scientist, June 8, 1998

Research in genetics has changed the way scientists view many disorders that befall patients. For example, investigators have taken giant steps in understanding the molecular basis of diseases such as cancer and cystic fibrosis. Genetic research also has radically revamped the understanding of afflictions — including manic-depression and obesity — that in the past were blamed on the infirmity and weak will of their sufferers.

Although genes now dominate many explanations of disease, more familiar units of human life — such as families and ethnic groups — play a major role in research. These two groups provide researchers with relatively uniform DNA samples, greatly streamlining searches for culprit genes.

The growing trend to focus research efforts in genetics on various groups identifiable by ancestry has forced many investigators and research institutions to take a fresh look at standards and practices for assuring the confidentiality of

research subjects. **Jeffrey R. Botkin,** a pediatrician at the University of Utah, recently completed a study examining methods and standards among investigators and journals in protecting the privacy and confidentiality of families involved in research. The study, appearing in the June 10 edition of *JAMA - The Journal of the American Medical Association*, promises to stir debate and discussion among researchers and editors. Although Botkin could not discuss specifics of the study prior to publication, he did offer that it points to "a definite need for discussion on how journals, investigators, and institutional review boards can develop consistent policies and practices that adequately protect the privacy of subjects in genetic research."

For the study, Borkin surveyed journal editors and researchers to assess their policies and practices with respect to confidentiality and privacy. While the identity of individuals is almost always withheld by investigators in reported results, pedigrees — standardized charts that display familial relationships and various phenotypes under consideration — are published with many articles. "In research literature, there has been sensitivity about publishing identifying data, base descriptions, and photos for individuals for quite some time." Borkin explains. "But genetic studies offer particularly interesting and problematic sets of information: the pedigree."

Even when individual family members are "anonymized," pedigrees can contain quite a bit of personal information on the genetic, medical, and reproductive histories of families. For example, when analyzed by a geneticist, a pedigree can reveal the mode of inheritance for a particular genotype under question. If an autosomal recessive gene is present in a family, the researcher can often conclude that a person is a carrier of the genotype even if the phenotype is not expressed. It also is possible that the identity of a family can be inferred from the chart, especially if the family structure is unusual or if a particularly rare disorder is being studies, according to Borkin.

"Pedigrees also can include histories of incest, misattributed parenthood, and abortion," Botkin says. "All of this sensitive information can show up here and the concern is that some people in society could decipher this information about you from medical literature. While employees and insurance companies typically don't read research literature and probably can't figure out the identity of subjects, some members of families can figure out identities."

In this survey, Borkin also found that researchers are divided over the best way to modify pedigrees to assure confidentiality. Researchers use two common practices: masking, changing the pedigree so that it is obvious to the reader that information is being withheld; and altering, modifying the pedigree by omitting information in such a way that is not obvious to the reader. Borkin found that although most professional societies in human genetics have taken a stand against altering research date, it is occasionally practiced by some researchers because it can offer greater assurance of confidentiality.

Cultural Privacy

To investigators, such as Janice A. Egeland, a professor of psychiatry and epidemiology at the University of Miami School of Medicine and a visiting professor of psychiatry at the University of Pennsylvania School of Medicine, assuring confidentiality not only has been important in published works, but has been a major concern in the day-to-day operations of the research center. For almost 40 years, Egeland has worked continuously with the Amish of Central Pennsylvania (S. Hoffert, The Scientist, 12[10]:1, May 11, 1998). She also has directed the Old Order Amish Affective Disorders Project at a University of Miami research office in Hershey, Pa., for more than 20 years.

She coordinates a comprehensive clinical and genetic investigation that includes collaborators from the National Institute of Mental Health, Albert Einstein College of Medicine, Case Western Reserve University, Columbia University, Yale University, and the University of California, San Diego. Researchers from the project changed the way psychiatrists think of affective disorders when they discovered DNA markers conferring a strong predisposition to manic-depressive disease (J.A. Egeland, et al., Nature, 325:783-87, 1987).

Egeland's efforts to assure the confidentiality and privacy of Amish patients and their culture started when she was working on her Ph.D. at Yale in the 1960s. While working on her dissertation exploring the relationship between belief and health, Egeland had become impressed with the kindness and candor the Amish showed in welcoming her into the community and helping with the research project. In the course of her studies, little-known and commonly misunderstood traditions among the Amish, such as bundling, powwowing, and folk-medicine practices, were revealed to her in some detail.

"There was significant encouragement from my colleagues to publish the dissertation," Egeland recalls. "Prior to my study, little was known about health and health care practices among the Amish. The view of many was that these people were a medical gold mine."

While many freshly minted Ph.D.'s would jump at the opportunity to publish their dissertation, Egeland went in the opposite direction. She decided to put the study on reserve at Yale and required researchers to submit letters explaining their need to see the study. "It was my view then and now that I didn't want this information to get into the hands of the wrong people," Egeland says. "The Amish had given me their trust and I wanted to protect them from the prying eyes of the outside world. In the popular press, their culture tends to be sensationalized. As a researcher, I had to take some responsibility for respecting their privacy as patients and a society."

Egeland's concern for protecting the confidentiality of patients and respecting the Amish culture has dominated how research is conducted at the Hershey project center. For example, with increasing frequency, she has had requests from outside investigators to help ease the collection of specimens among the Amish. While her duties as principal investigator on the affective disorders project limit her ability to sponsor these studies, she is most concerned about how some proposals and requests show a lack of respect for the Amish culture and their values.

"I am just amazed and shocked by some of the requests I get." Egeland says. "Some researchers seem to be willing to go around accepted practices of informed consent when obtaining samples. Others make little effort to make sure that their research does not conflict with the culture and moral sensibilities of the Amish. It's come to the point where I don't have time to consider any work with investigators outside my network."

Reconsidering Consent

Concern over the potential of genetic research to stigmatize groups of research subjects also has stimulated the search for ways to supplement the process of informed consent. Since the 1960s, individual informed consent has served as the cornerstone of research practices designed to protect human subjects. But Morris W. Foster, an associate professor of anthropology at the University of Oklahoma, argues that advances in genetic technology have made it necessary for researchers to engage whole communities in discussion about research plans. Because genetic research can be misused, community discourse is needed to assure that individuals can make truly informed decisions.

"In publishing their results, researchers routinely identify research participants by ethnicity and geographic origin and open them up to stigmatization and adverse perceptions from society," Foster says. "To be sure, scientific information itself is not the source of this problem. But the reality is that lay people can use this information in ways that can hurt research subjects. This fact extends the ethical obligations of researchers."

Although Foster does not suggest using communal discourse as a replacement for individual consent, he says discourse with the community is essential to assure that people can make informed decisions about participating in research. Frequently, in research involving Native American communities, Foster says it is necessary to engage extended families and even whole tribes in discussion about ramifications from scientific investigation. Public meeting were initiated during which researchers sought out representatives from various social units. When specific concerns about the research project were raised, further discussion of research plans was needed, and terms of the project were renegotiated.

Botkin agrees that community discourse often is required for subjects to make informed decisions. But he adds that where stigmatization is a concern, researchers might consider not identifying research subjects as members of a particular group. "While the issue would have to be considered on a case by case basis, I don't have any problem withholding information that identifies the research subjects as members of any social group when the information is not relevant to the results," Borkin says. But he is quick to note that deciding relevance is difficult even if research subjects face possible disgrace.

Some discussions among researchers in the field have focused on how investigators with a bona fide need to know the identity of groups studied can obtain this information if it doesn't appear in the original publication. Just how this information could be made available to investigators is uncertain, according to Borkin: "Journals have been clear that they don't want to take on this function. But I can envision something like the Negative Clinical Trials eventually serving this purpose in genetic research. ❑

* * * * * * * * * * * * * * * * * * * *

Questions:

1. What are the two common ways that researchers try to maintain the confidentiality of family histories?

2. Why is the community's involvement in genetic research so important?

3. What advantages can you think of to studying an isolated population such as the Icelandics of a group of people that tend to only breed with people from the same ancestry?

Answers are at the end of the book.

31

Animal behavior is the link between organisms and environment and between the nervous system and the ecosystem. Behavior is one of the most important properties of animal life. It plays a critical role in biological adaptations and is how we as humans define our own lives. The comparative study of behavior over a wide range of species can provide insights into influences affecting human behavior. For example, the woolly spider monkey in Brazil displays no overt aggressive behavior among group members. We might learn how to minimize human aggression if we understood how this species of monkey avoids aggression. If we want to have human fathers be more involved in infant care, we can study the conditions under which paternal care has appeared in other species like the California mouse or in marmosets. Studies of various models of the modes of communication in birds and mammals have had direct influence on the development of theories and the research directions in the study of child language. Research by de Waal on chimpanzees and monkeys has illustrated the importance of cooperation and reconciliation in social groups. This work provides new perspectives by which to view and ameliorate aggressive behavior among human beings.

The Evolution of Morality

by Kim A. McDonald

The Chronicle of Higher Education, March 15, 1996

If chimpanzees learn anything at the Arnhem Zoo in the Netherlands, it's to be on time for dinner.

Zookeepers will not feed any of them until all of the chimps have left the outdoor enclosure and entered their sleeping quarters. So the chimps don't dally.

Their hunger, however, is not the only reason for the rush indoors. The chimps adhere to a strictly enforced rule not to keep others in the colony waiting.

What can happen to violators was demonstrated when two adolescent females decided to remain outside for two hours one balmy evening, while the rest of the chimps waited to be fed.

Zookeepers, fearing that a fight might break out, separated the tardy individuals from their hungry companions that night. But the colony didn't forget the transgression.

When all of the chimps were released together the next morning, recalls Frans B.M. de Waal, the other chimps "gave a terrible beating to those two females — who, that same evening, came in first."

Rules and Expectations

The notion that chimpanzees have rules of conduct and expectations of others within their societies comes as no surprise to Mr. de Waal, a primatologist who is a professor of psychology at Emory University.

For more than two decades, his research at the Arnhem Zoo, at the University of Wisconsin's primate center, and now at Emory's Yerkes Regional Primate Center in Atlanta, has shown that many behavioral traits once assumed to be uniquely human are actually shared by other primates (*The Chronicle*, July 7, 1995).

In his first book, *Chimpanzee Politics: Power and Sex among Apes* (Johns Hopkins University

Press, 1982), Mr. De Waal's descriptions of the soap-opera-like antics of chimpanzees at the Arnhem Zoo show how these primates formed alliances with one another to displace stronger rivals or to plot against others on whom they sought revenge.

Next in *Peacemaking among Primates* (Harvard University Press, 1989), he used the results of his research on food sharing and conflict resolution in a variety of primates to argue that sharing, peacemaking, and forgiveness are not cultural inventions, but integral parts of our primate heritage.

His most recent work, *Good Natured: The Origins of Right and Wrong in Humans and Other Animals*, to be released by the Harvard Press this week, takes his ideas about the evolutionary basis of human behavior one step further — to a re-examination of our concepts of justice and morality.

Mr. de Waal argues in his book that our sense of justice and the framework of our moral behavior are innate traits, and that the underpinnings of both are evident in higher primates and other social animals.

He is not the first scientist to suggest that moral or ethical behavior has a biological basis. However, experts in ethology, the study of animal behavior, say his book is the first to make a comprehensive argument, using examples from throughout the animal kingdom, about how those behaviors may have evolved from animals to humans.

'Breaking New Ground'

"He's presented a very convincing case for continuity," says Barbara B. Smuts, a professor of psychology and anthropology at the University of Michigan, who studies baboons.

"It's breaking new ground," says Meredith F. Small, an associate professor of anthropology at Cornell University, who studies macaques. "He's essentially coming out with a hypothesis: that you can find the basis of morality and justice in other animals. And I think he's going to be taken seriously."

Not everyone in the social sciences is likely to agree, however. Many traditionalists still regard justice and morality as human inventions that serve to control individuals in our society. It's a view of human nature that Mr. de Waal calls "very Calvinist — that human nature is all bad, and we must work very hard to overcome it."

But if our sense of morality is innate, a product of our evolution, can other animals be considered moral? Or must an animal be human, by definition, in order to be humane?

Mr. de Waal concedes that it's the latter, if only because we have insisted on attaching anthropocentric labels to charitable or moral behavior. In fact, his book provides many examples of social animals that behave in ways humans might call "ethical": whales and dolphins that risk their lives to save injured companions, chimpanzees that come to the aid of victims of aggression, elephants that attempt to revive slain comrades and then refuse to leave them. But he stops short of labeling these animals moral creatures.

"I'm not claiming that animals have morality," he says. "I'm claiming that animals have many of the emotions and psychological mechanisms that are involved in human morality. I don't think they piece it together exactly as we do. There are some differences in human morality and what we see in animals."

Sympathy and Empathy

There are also many similarities, and they are at the core of his argument. Mr. de Waal contends that without the evolution of sympathy and empathy, moral behavior would be impossible. "They are very much pillars of human morality," he says.

Sympathetic behavior, he says, is seen in a variety of higher mammals: dogs that stay protectively close to crying children; monkeys that help handicapped individuals; killer whales that follow their sick companions into shallow water and beach themselves with them.

In contrast, empathy — the ability to comprehend the feelings and thoughts of others — is reserved for species of high cognition.

This is so, psychologists believe, because

empathy first requires an individual to distinguish oneself from another. Such self-awareness is often defined in children and animals as the ability to distinguish one's image and notice such changes as painted dots on one's face.

Consoling Victims

Besides humans over the age of 18 months, only two other species are known to have that ability — chimpanzees and orangutans. Chimpanzees, Mr. de Waal says, regularly display empathy toward one another by "approaching victims of aggression and consoling them by putting their arm around them."

Like us, chimpanzees are also masters of reciprocity, trading food and other favors in the expectation of receiving something in return. That sense is so well developed in chimps that zookeepers who leave brooms or other articles in their cages need only hold up a piece of fruit and point at the forgotten article to get them to retrieve it.

As in human societies, in which "one good turn deserves another," chimps who are recipients of favors and fail to reciprocate produce feelings of indignation in others. Mr. de Waal found in his studies that chimps who share food willingly with others are most likely to receive food in return, while those who don't share rarely receive anything.

Expression of Indignation

In fact, when these selfish individuals beg, they become objects of aggression — an expression of indignation that, he believes, is at the root of our sense of justice. "When animals feel they owe someone else something, the concept of fairness is not far behind."

This tit-for-tat system among chimpanzees and other apes extends to alliances that are formed to defeat stronger rivals, to extract revenge against others, or to punish violators of codes of ethics, such as a young male monkey caught mating in the presence of a dominant male, or the female chimps at the Arnhem Zoo who were late for dinner.

"Non-human primates and some other animals follow certain rules of conduct, and they enforce those rules by punishing transgressions," says Mr. de Waal. "And they even have expectations of one another that go beyond the hierarchy and relate to reciprocity and to mutual obligation."

He believes that these behaviors, combined with empathy and sympathy, form the elements of what evolved into our sense of fairness, justice, and morality.

"The fact that the human moral sense goes so far back in evolutionary history that other species show signs of it plants morality firmly near the center of our much-maligned nature," he wrote. "It is neither a recent innovation nor a thin layer that covers a beastly and selfish makeup."

Supposedly Selfless Acts

If his book persuades other social scientists that this is the case, our philosophical views of morality could be changed dramatically. In fact, Mr. de Waal believes we are now reaching the point at which biology may be finally able to "wrest morality from the hands of philosophers."

But he concedes that some scientists will find it difficult to accept that traits promoting good-natured behavior could have arisen through natural selection, which biologists have come to view as a harsh process, in which individuals are manipulated by their genes to cheat, kill, and compete with others in their group to propagate their genes as widely as possible.

Some scientists see no inconsistency, arguing that most supposedly selfless acts, such as giving blood, benefit individuals by raising their status within society.

"It's not a problem at all if you see moral systems as a system of contracts," says Richard D. Alexander, a professor of evolutionary biology at Michigan and author of *The Biology of Moral Systems* (Aldine, 1987). "And humans are restricted to doing those things in which the cost-benefit rule is on the benefit side."

'Keeping the Group Harmonious'

But Mr. de Waal, who has built on Mr. Alexander's work, believes that we also have traits that are fundamentally good. The reason social animals live in groups, he says, is for the benefits of group life, such as protection from predators. "So, there's an interest in keeping the group harmonious," he says. The traits promoting moral behavior and fairness within groups, as well as the tendency to demonstrate what he calls "community concern," serve to do just that.

"Natural selection is harsh, and natural selection is selfish. But the product that is delivered doesn't need to be that way," he says. "We know that natural selection has produced symbiosis, cooperation, and sharing responses — and most likely natural selection has produced human morality as well.

"Not the specifics of it," he adds, "but certainly the general outline." ❏

* * * * * * * * * * * * * * * * * * * *

Questions:

1. How do traditionalists view justice and morality?

2. What does Mr. de Waal contend to be the core of moral behavior?

3. According to Mr. de Waal, why do social animals live in groups?

Answers are at the end of the book.

- 32 -

Medical genetics is a fascinating and swiftly evolving subspecialty, spanning the complete path of human development from reproduction through the prenatal, neonatal, pediatric, adolescent, and adult periods. The Human Genome Project, designed to sequence the entire human genome by the year 2005, will undoubtably alter the practice of medicine. Screening protocols, early intervention, and new treatment modalities, including gene therapy, will become the standard of care for familial cancers, inborn errors of metabolism, and common adult-onset disorders such as diabetes, Parkinson's, and heart disease, to name just a few. It is imperative for physicians today to understand basic genetic, cytogenetic, and molecular genetic concepts.

Making Geneticists Out of Physicians

by Adam Marcus

Nature Medicine, June 6, 1996

Whether it is a sign of the times or a watershed in health care, all doctors may soon be required to demonstrate proficiency in interpreting genetic tests and explaining the results to their patients. A Task Force on Genetic Testing, jointly sponsored by the US National Institutes of Health (NIH) and the Department of Energy (DOE), which met last month in Baltimore, says genetics needs to become an essential element of medical practice.

The task force's opinion reflects growing consensus among medical policy makers that physicians, regardless of specialty, will confront genetic testing at some point in their careers.

"We think [all physicians] should know something about genetics," said Neil Holtzman, the task force chairman. "To ensure that they had it, they would need some demonstration of competency in genetics and genetic testing for licensure or relicensure."

However, Holtzman, who is also director of the Johns Hopkins Medical Institutions' center for Genetics and Public Policy Studies, cautioned that the licensing issue is still in its infancy. "We're just beginning to discuss" it, he said. Even so, he said, "Somewhat surprisingly, the task force seemed to be in agreement."

The task force was established in 1994 by the NIH-DOE Working Group on Ethical, Legal, and Social Implications of Human Genome Research, or ELSI, and consists of 15 voting members from the private and public sectors, including insurance industry, biotechnology and professional society representatives. In addition, there are five non-voting government representatives on the committee.

Holtzman said the recommendation that physicians be required to demonstrate competency in genetics and genetic testing — a yet-to-be defined standard — was supported by the American Medical Association's (AMA) representative to the task force, Patricia Numann. In addition, Myron Genel, another member of the AMA's Council on Scientific Affairs, agreed with the recommendation. "The explosion of knowledge is such that there are certainly not enough geneticists or genetic counselors" to handle the growing interest of patients in the technology, Genel says. "Physicians are going to need to become more versed in the applications and limitations of genetic testing."

Genel, who is also a professor of pediatrics and associate dean at the Yale University School of Medicine, indicated that the position taken by the AMA at the Baltimore meeting was not the official stance of the association at large, and would not be put to a vote until next December, at the earliest. However, he said the council has been considering the implications of genetics for AMA members for some time.

Holtzman was involved in a 1991 survey of selected primary care physicians and psychiatrists to determine how well-versed these doctors were in genetics. The study, published in the August 1993 edition of *Academic Medicine,* found that those doctors who responded answered fewer than three-quarters of the genetics questions correctly. Professional geneticists in the survey, by comparison, answered nearly 95 percent of those questions correctly.

Thus, the study's authors concluded, "curriculum planners should ensure that instruction in genetics (including relevant skills such as the interpretation of probabilistic results and their explanation to patients) is integrated throughout all years of medical school and is provided to residents in primary care."

Janet Hafler, associate director of faculty development at Harvard Medical School, said that should the interim principles eventually become official guidelines, Harvard would adapt. "Schools do change to abide by guidelines," said Hafler, who noted that Harvard and other institutions moved swiftly to incorporate HIV into their core curricula.

Holtzman said the task force will convene again in September, and should have a final report by the end of this year. At that time, were the report to include some incarnation of a competency requirement, implementation could require action by such bodies as the Food and Drug Administration, state or federal legislators, and professional societies. ❏

* * * * * * * * * * * * * * * * * * * *

Questions:

1. Why do physicians need to understand genetics and genetic testing?

2. What does ELSI stand for?

3. What were the results of Holtzman's 1991 study of how well-versed doctors were in genetics?

Answers are at the end of the book.

- 33 -

Anyone who has watched a wrinkle slowly gouge his or her face like a strip mine, or has been disturbed by a loss of memory, has uncomfortably confronted the human aging process. The inexorable march of time on our bodies begs an important question: why do we have to grow old? Theories of aging fall into two categories — (1) wear-and-tear theories and (2) programmed aging theories. The wear-and-tear theories describe aging as an accumulation of damage and waste that eventually overwhelms our ability to function. The programmed theories propose a clock in our bodies that controls not only our process of development but also triggers our self-destruction. Two of the wear-and-tear theories are the somatic mutation theory and the free radical theory. The somatic mutation theory holds that (wear-and-tear) errors in the DNA accumulate so that eventually the proteins it makes no longer work. The free radical theory of aging proposes that highly-reactive atoms or molecules attach to other molecules, turning them into free radicals, which in turn attack other molecules, and so on, thereby inactivating a great number of molecules. The free radical theory has the advantage that it seems to be able to encompass the other wear-and-tear theories of aging. But, even though wear-and-tear definitely does occur, the patterns of aging that we see do not completely fit that model. Life-spans are quite species-specific; rats live just a few years whereas Galapagos turtles live over a hundred. The genetic program is thus an important factor in aging. The identification of "Age" mutations in the worm Caenorhabditis elegans *has placed a greater emphasis on the actions of stress-response genes in prolonging life-span rather than on the action of clock-type genes in shortening it.*

Mechanisms and Evolution of Aging

by Gordon J. Lithgow and Thomas B. L. Kirkwood

Science, July 5, 1996

Single-gene mutations that extend the life-span of the worm *Caenorhabditis elegans* dramatically demonstrate the genetic basis of aging and may eventually lead to the elucidation of aging mechanisms. At the same time, evolution theory provides a powerful insight into the genetic basis of aging, and recent experiments are allowing these ideas to be tested and refined.

One of the clearest messages to emerge from evolution theory is that aging is not likely to be regulated in the same programmed way as are early life processes such as development (*1*). The reason is simple. In the natural world, organisms die from a wide array of extrinsic as well as intrinsic causes, and in most species, few individuals live long enough to show obvious signs of senescence. Natural selection cannot plausibly explain the evolution of a temporal control system or "aging clock" whose primary function is to bring about senescence and death in only a handful of survivors, especially when this action is detrimental to the individuals in which it occurs.

Nevertheless, it is abundantly clear that genes do influence aging and longevity (*2*). What kinds of genes are these likely to be? Some of the strongest candidates are genes that regulate the process of somatic maintenance and repair, such as the stress-response systems. Maintenance is beneficial and necessary up to a point, but most maintenance systems have hidden costs, for example, an energy requirement. The disposable soma theory of aging (*3*) points out that it is actually disadvantageous to increase maintenance beyond a level sufficient to keep the organism in good shape through its natural life expectancy in the wild, because the extra costs will eat into resources that in terms of natural selection are better used to boost other functions that will enhance fitness.

With its short 20-day life-span and accommodating genetics, the microscopic *C. elegans* worm is fast becoming an important system for genetic studies on aging. Are the life-span extension mutants of *C. elegans* consistent with the evolutionary theories of aging? The first "Age" mutation to be described, age-*1*(*hx546*), increases mean life-span by 65% and maximum life-span by 110% (*4*). Under normal laboratory conditions, strains carrying age-*1*(*hx546*) appear very similar to wild-type strains in their length of embryogenesis, postembryonic developmental time, and fertility period, with some slight reduction in reproductive output (*4, 5*). How can a single-gene mutation have such a dramatic effect on the rate of aging? One plausible model is that *age-1* is a coordinate regulator of a range of stress-response genes that shift cellular physiology toward maintenance. This model is based on the finding that age-1 mutant strains are better equipped than their wild-type counterparts to overcome the effects of extrinsic stress (*5-9*). First, age-*1*(*hx546*) confers resistance to hydrogen peroxide (H_2O_2) and paraquat (*6, 7*). Both of these agents cause oxidative stress by promoting the generation of highly reactive hydroxyl radicals. The mutant also accumulates fewer deletions of the mitochondrial genome, an age-related phenomenon thought to be the result of damage by radicals (*10*). The idea that oxidative stress resistance is due to a regulatory alteration comes from the observation that *age-1* mutants have elevated activities of the antioxidant enzymes Cu,Zn-superoxide dismutase and catalase, which together detoxify H_2O_2 and paraquat (*6, 7*).

The *age-1* mutant was also found to have increased intrinsic thermotolerance. When young worms are subjected to lethal heat shocks, the

age-1 mutants are 40% more resistant than are their wild-type counterparts (*5*). This finding, together with the fact that short, nonlethal heat shocks induce both thermotolerance and extended life-span, leads to the suggestion that heat shock proteins may slow aging processes (*8*). Finally, *age-1* mutant worms are resistant to ultraviolet (UV) radiation, a phenotype that has proved an excellent predictor of life-span (*9*).

Critical support for the idea that stress resistance is causally related to extended life-span comes from the analysis of other Age mutants (*11-14*). Mutations in the genes *daf-2, daf-23, daf-28,* and *spe-26* extend life-span and also confer a set of stress resistance phenotypes. All the Age mutations tested so far are resistant to oxidative stress, thermal stress, and UV radiation — with the degree of resistance often proportional to the extension in life-span. It would be parsimonious to conclude that these mutations extend life-span by a common mechanism. Classical genetic analysis of these mutations lends support to this notion. The life-span phenotypes of *daf-2, daf-23, spe-26,* and *age-1* all depend on the wild-type function of another gene, *daf-16* (*9, 11, 13*). The *daf* genes are well known to nematode geneticists because they encode components of a signal transduction pathway critical during development. It now appears that this pathway influences the maintenance of the adult worm.

An apparent challenge to the evolutionary arguments against an aging clock is the discovery of a new class of *age* genes defined by the *clk* mutants (for "abnormal function of biological clocks"). The *clk* mutants display a range of phenotypes — including slow development, a slow cell cycle, slow and irregular behavioral rhythms, and extended life-span (*14*). The mutants seem to have lost temporal control, but even *clk-1* confers resistance to UV light, so these mutations may eventually fall into line with the other Age mutations (*9*).

It is too soon to know whether the Age mutations, and genetic analysis in general, will validate the carefully prepared evolutionary theory of aging. However, we can be encouraged by the emerging story, which places an emphasis on the

155

actions of stress-response genes in prolonging life-span rather than on the action of clock-type genes in truncating it. ❑

References

1. T.B.L. Kirkwood and T. Cremer, *Hum. Genet.* **60**, 101 (1982).
2. C.E. Finch, *Longevity, Senescence, and the Genome* (Univ. of Chicago Press, Chicago, 1990).
3. T.B.L. Kirkwood and C. Franceschi, *Ann. N.Y. Acad. Sci.* **663**, 412 (1992).
4. D.B. Friedman and T.E. Johnson, *Genetics* **118**, 75 (1988).
5. G.J. Lithgow, T.M. White, D.A. Hinerfeld, T.E. Johnson, *J. Gerontol. Biol. Sci.* **49**, B270 (1994).
6. P.L. Larsen, *Proc. Natl. Acad. Sci. U.S.A.* **90**, 8905 (1993).
7. J.R. Vanfleteren, *Biochem. J.* **292**, 605 (1993).
8. G.J. Lithgow, T.M. White, S. Melov, T.E. Johnson, *Proc. Natl. Acad. Sci. U.S.A.* **92**, 7540 (1995).
9. S. Murakami and T.E. Johnson, *Genetics*, in press.
10. S. Melov, G.J. Lithgow, D.R. Fischer, P.M. Tedesco, T.E. Johnson, *Nucleic Acids Res.* **23**, 1419 (1995).
11. C. Kenyon, J. Chang, E. Gensch, A. Rudner, R. Tabtiang, *Nature* **366**, 461 (1993).
12. W.A. Van Voorhies, *ibid.* **360**, 456 (1992).
13. P.L. Larsen, P.S. Albert, D.L. Riddle, *Genetics* **139**, 1567 (1995).
14. A. Wong, P. Boutis, S. Hekimi, *ibid.* p. 1247.

* * * * * * * * * * * * * * * * * * * *

Questions:

1. What genes are most likely involved in aging and longevity?

2. What do all the "Age" mutations tested so far have in common?

3. Describe the phenotype of the *clk* mutants.

Answers are at the end of the book.

Answers

to

End of Article

Questions

ANSWERS TO END OF ARTICLE QUESTIONS

PART ONE: GENETIC ANALYSIS

- 1 - Mendelian Proportions in a Mixed Population

1. Population equilibrium that is established through random mating depends on the existing distribution of allele frequencies and also that the allele frequencies remain constant.

2. p = pure dominant (AA)
    2q = heterozygote (Aa)
    r = pure recessive (aa)

3. Large populations help increase the chance for random mating and make the short-term effects of migration, selection and mutation negligible.

- 2 - Privatizing the Human Genome

1. Venter's company will be using faster, more automated sequencing machines and a different sequencing strategy, making the target date of 2001 very feasible.

2. One hundred million base pairs a day.

3. The whole-genome shotgun approach has only been used on much simpler organisms and results in some gaps of information. This approach does not meet the goal of developing a gold standard representation of the genome — largely continuous, with few if any gaps.

- 3 - The Politics of Germline Therapy

1. The Convention's principles permit preventive, diagnostic or therapeutic intervention in the human genome only if its aim is not to introduce any modification in the genome of descendants which would appear to bar germline therapy.

2. The California anti-cloning law's broad definition of cloning inadvertently banded oocyte nuclear transfer, which is not cloning at all.

3. This procedure could be used to avoid a mitrochondrial disorder that is present in the patient's family.

- 4 - Bearing False Witness

1. In the order of a million to one.

2. Between 1 in 114 to 1 in 468.

3. Britain's Forensic Science Service uses mtDNA only to eliminate suspects or to back up other pieces of evidence.

4. Juries need to be told that a random match is far more likely to occur with mtDNA than with nuclear DNA analysis.

- 5 - Have You Used an Adeno Vector...Lately?

1. In order for gene delivering methods to be effective in gene therapy, proper expression of the transduced gene in the right cells at the right time and at the appropriate level of expression needs to be met.

2. Homologous recombination would have an advantage over non-homologous recombination strategies in gene therapy when performing experiments involved in the replacement of dominant-acting mutations and in the insertion of a suicide gene.

3. To detect homologous or targeted integration events, they first infected HeLa cells harboring a defective neomycin-resistance (*neo*) gene with an adeno-associated virus vector harboring a different insertion mutation in the same gene. Homologous recombination between the two mutated genes would generate a functional gene and lead to neomycin resistance.

- 6 - The Commercialization of Human Genetics; Profits and Problems

1. According to studies by Blumenthal *et al.*, the most cited reasons for delaying publication of research results is for filing a patent application.

2. Seventy-six percent of all human DNA sequence patents are owned by private industries, 17% are held by public groups and 7% by individuals.

3. Healy argues that it is too early to use BRCA gene testing in everyday clinical practice because it violates a common-sense rule of medicine: don't order a test if you lack the facts to know how to interpret the results.

PART TWO: THE CHEMISTRY OF INHERITANCE

- 7 -   Molecular Structure of Nucleic Acids: A Structure for Deoxyribose Nucleic Acid

1. The Pauling and Corey model consists of three intertwined chains, with the phosphates near the fiber axis and the bases on the outside. The Watson and Crick model suggests the bases are inside the helix and the phosphates are on the outside. Their structure has two helical chains each coiled around the same axis.

2. The two chains are held together by the purine and pyrimidine bases.

3. The extra oxygen atom in the ribose sugar would make too close a van der Waals contact between the two chains.

- 8 -   Histone-GFP Fusion Protein Enables Sensitive Analysis of Chromosome Dynamics in Living Mammalian Cells

1. True. Since double minute chromosomes lack a functional centromere.

2. Double minute chromosomes contain a diversity of amplified oncogenes and undergo uneven segregation and accumulation causing increase malignant potential during tumor formation.

3. Double minute chromosomes display a cluster behavior and have a very close association with normal chromosomes, which may provide one mechanism by which they are transmitted to daughter cells despite lacking a functional centromere.

- 9 -   Genetical Implications of the Structure of Deoxyribonucleic Acid

1. Adenine pairs with thymine and guanine pairs with cytosine.

2. Other researchers using X-ray data have shown similar patterns for the structure of DNA from both isolated fibers and certain intact biological material.

3. Spontaneous mutations may be due to a base occasionally pairing with its complementary base (for example, adenine pairing with cytosine).

- 10 - Far from the Maddening Cows

1. Yeast normally reproduce by budding; yeast have 16 chromosomes.

2. Many scientists argue that DNA or its RNA counterpart contained in microorganisms likes bacteria and viruses can only carry infections.

3. False. New research on prions demonstrates that proteins are flexible and can flip from one state to another.

4. Imprinting is a form of inheritance that allows genes to behave differently depending on whether they are inherited from the mother or the father.

- 11 - A Long Molecular March

1. The human genome is estimated to be 3 billion base pairs long.

2. Only about 3-5% of the genome is actually used by cells to make proteins.

3. The human-genome project (1) seeks to support medical research by gathering genetic information about humans and other organisms; (2) is establishing physical maps so researchers can obtain a piece of DNA they want from any portion of the chromosomes; and (3) is sequencing the entire human genome.

- 12 - We Ask They Answer

1. Despite the fact that most of the proteins of the egg cytoplasm are unidentified, their role is to prepare the sperm's nucleus for its union with the egg's genetic material and to orchestrate the first cell divisions that turn the embryo into a ball of 8 or 16 cells.

2. Theory #1 — adult cells are more set in their ways than cells of an embryo that are easier to clone.
Theory #2 — adult cells are damaged beyond the egg's ability to repair them.

3. Pluripotent is the capability of a cell to generate into any type of adult cell.

PART THREE: CHROMOSOMES: STRUCTURE, FUNCTION, AND REPLICATION

- 13 - The Great Divide

1. Dividing chromosomes split at the centromere.

2. False. The DNA sequence of centromere varies widely — not only between species but also between single organism's chromosomes.

3. Researchers suspect the tip was passed on to daughter cells through cell division.

- 14 - One FISH, two FISH, red FISH, blue FISH

1. The biggest obstacle for the utilization of FISH analysis was its inability to distinguish more than one target sequence.

2. This abnormality was not detectable by banding pattern analysis because the banding pattern of the involved chromosome bands and the size of the translocated material was so similar.

3. (1) Cost; (2) sensitivity and specificity not established; (3) regulatory agencies have not approved FISH analysis for clinical use; (4) complete cytogenetic analysis needed despite FISH analysis due to probe limitations; (5) low sample volume; and (6) perception that FISH is a "molecular" test.

PART FOUR: GENE LINKAGE AND CHROMOSOME MAPPING

- 15 - Fatal Familial Insomnia

1. The mutant allele at codon 129 of the prion protein gene *PRNP* encodes for methionine in FFI and valine in Creutzfeldt-Jakob disease.

2. The patient was homozygous at codon 129 but showed clinical features commonly seen in heterozygotes, such as myoclonic jerks, cerebellar signs, and intellectual deterioration, pseudoperiodic triphasic slow waves at EEG in more advanced stages of the disease and nerve cell loss, gliosis, and focal microspongiosis in cerebral cortex.

3. Extensive thalamic degeneration may be the pathologic correlate of the dysfunction of circadian rhythms in FFI patients.

- 16 - Dyskeratosis and Ribosomal Rebellion

1. A typical appearance of a patient with dyskeratosis congenita is little hair, almost no eyebrows, evanescent nails, teeth in poor condition, patches of discoloration on his skin and sometimes tearful due to an abnormality of the lacrimal ducts.

2. The exact role of the *DKC1* protein product is uncertain but it seems likely that it is a nucleolar protein and responsible for early steps in rRNA processing.

3. The link between dyskeratosis congenita and malignancy is serious because there is a certain phenotype caused by an inherited mutation and there is the threat imposed by the uncertain stochastic nature of superimposed somatic mutations.

- 17 - Mutations in the *parkin* Gene Cause Autosomal Recessive Juvenile Parkinsonism

1.
AR-JP	Parkinson's Disease
-autosomal recessive	-complex inheritance; except for rare autosomal dominant families
-no Lewy bodies	-Lewy bodies
-mutations are deletions	-missense mutations

2. False. Mutations in α-synuclein are only associated with Parkinson's disease and are not seen in AR-JP.

3. Parkin lacks a lysine at position 63 and the carboxy-terminal glycine at position 76 that are essential for ubiquitination.

- 18 - Alopecia Universalis Associated with a Mutation in the Human *hairless* Gene

1. FGF5 is an inhibitor of hair elongation in mice.

2. The most common form of hair loss is androgenetic alopecia (male pattern baldness) and affects about 80% of the population.

3. All patients of the Pakistani pedigree showed a homozygous A-to-G transition, resulting in a missense mutation converting a threonine to an alanine at amino acid residue 1022 of the human *hairless* protein.

4. To identify the alopecia universalis locus segregating in the Pakistani family, the investigators initiated a genome-wide search for linkage by homozygosity mapping.

PART FIVE: REGULATION OF GENE EXPRESSION

- 19 - A Head Full of Hope

1. Suicide genes are injected into target cells to cause them to self-destruct or kill themselves.

2. Viruses serve as vectors or delivery vehicles that carry the therapeutic gene to the target cell.

3. The majority of the approved gene therapy techniques are divided into two categories: those that are aimed at correcting defective genes and those that are involved in inducing specific cells to produce proteins that would make them vulnerable to attack by the immune system or drugs.

4. Glioblastomas kill specific areas of the brain and do not spread to other parts of the body like other forms of cancer. Therefore, a treatment would only have to work locally, not systemically. In addition, since brain cells do not divide, the virus carrying the gene would infect only rapidly dividing cells that would be the cancer cells. Lastly, the brain's immune system is weak and injections of the gene would not elicit an immune response like other parts of the body would.

- 20 - Effect of Age at Onset and Parental Disease Status on Sibling Risks for MS

1. The present study carefully examined the following covariates: sex of index case, sex of the siblings, birth cohort of the sibling, age of onset of MS in index case, and MS disease status of parents.

2. True. MS is seen at a lower rate in males than in females.

3. The sex of the sibling, parental MS status, and index patient onset age were the three important factors influencing MS risks to siblings.

4. Currently, three genome screens using linkage analysis have revealed that only the HLA region on chromosome 6p is consistently linked to MS susceptibility.

- 21 - Human Gene for Physical Performance

1. True. Previous studies suggested that the D allele was associated with a positive effect on physical performance. However, in this study the I allele was associated with improved endurance.

2. Duration of exercise improved significantly for those of II and ID genotype but not for the DD genotype. Improvement was 11-fold greater for those of II than for those of DD genotype.

3. The genotype-dependent improvements were most likely due to an improvement in the endurance characteristics of the tested muscles.

PART SIX: MUTATION, RECOMBINATION, AND REPAIR

- 22 - How to Gain the Benefits of Sexual Reproduction without Paying the Costs: A Worm Shows the Way

1. Three advantages to sexual reproduction is that genetic recombination and cross fertilization call the species to (1) bring together mutations arising in different individuals, (2) generate genetic variability and thus adapt to changing environments, and (3) shuffle their genes in every generation and thus keep parasites at bay.

2. The X chromosome of the male can be described as selfish because it enhances its own survival at the expense of other chromosomes (present in the nullo-X chromosome) in the same individual.

3. A haplodiploid life cycle — as seen in the parasitic wasp *Nasonia vitripennis* — is one where males develop from unfertilized eggs and are haploid and females develop from fertilized eggs and are diploid.

- 23 - Papillomavirus and p53

1. p53 is one of the most important cellular proteins in guarding repair processes and maintaining chromosome stability.

2. False. According to this study, having an arginine residue at amino acid position 72 makes one more susceptible to degradation by the HPV E6 protein.

3. The evidence that multiple, progressive cellular changes need to occur before a malignancy develops includes: (1) the long latency for tumor development after primary infection, (2) the observed monoclonality of anogenital tumors that contain HPV, and (3) the absence of tumor-specific modifications in the viral oncogenes.

- 24 - Genetic Testing: Out of the Bottle

1. Breast, ovarian, colon, and prostate. Lifetime risk for colon and prostate risk is only slightly higher than the general population risk, whereas breast cancer risk is 85% and ovarian cancer risk is from 40-60%.

2. Women most appropriate for testing would include those with a strong family history of early onset (pre-menopausal) breast and ovarian cancer and those with at least two or more close relatives with breast cancer. Genetic testing should be conducted within a framework of intensive counseling.

3. Fear of discrimination by employers and insurance companies who might learn of positive results.

PART SEVEN: THE GENETICS OF EVOLUTION

- 25 - Let There Be Life

1. Most ribozymes are as likely to snip an RNA molecule apart as stitch one together, which makes copying long molecules a difficult task.

2. Huge amounts of adenine and guanine were created as part of chemical reactions in Miller's experiment.

3. According to the complexity theory, when a system reaches some critical level of chaos, it naturally generates a degree of complex order. The primordial soup composed of nucleotides, lipids, and amino acids instantly became integrated into an organized system due to being part of a complex and chaotic mixture.

- 26 - Molecular Coproscopy: Dung and Diet of the Extinct Ground Sloth *Nothrotheriops shastensis*

1. PCR allows identification of the species or individual from which the droppings originated from as well as their diet and parasitic load of the animal.

2. Coprolite is ancient feces.

3. N-phenacylthiazolium bromide (PTB) is a reagent that cleaves glucose-derived protein cross-links to release DNA that might be trapped with sugar derived condensation products.

- 27 - The Human Genome Diversity Project: Medical Benefits Versus Ethical Concerns

1. The ethical guidelines were: (1) to obtain informed consent before collecting samples; (2) to respect the privacy of individual's samples; (3) to encourage the participation of the population in the development of study designs; and (4) to keep them fully informed of the results of various studies.

2. Pima tribe in Arizona, the Old Order Amish, The 'Gullah' Islanders in the American Southeast, and some West Africans.

3. Many populations in the world were horrified at the notion of having their DNA collected and studied because many of these groups believe that all aspects of their own persons — such as blood, hair and tissue, as well as DNA samples — are sacred.

- 28 - Tracking the History of the Genetic Code

1. DNA is translated into RNA and the RNA is read by cellular machinery to make an amino acid which in turn produces a protein.

2. Arginine

3. Freeland discovered that bases fall into two size classes, and mutations that swap bases of similar size are more common than mutations that switch base sizes. Secondly, during protein synthesis, the first and third members of a codon are much more likely to be misread than the second one.

- 29 - 'Lamarckian' Mechanisms in Darwinian Evolution

1. The current version of biological Darwinism is incomplete because it does not address, both, the evolution of new systems for acquiring, storing, and transferring information and the evolutionary effects of such systems once they are in place. Suicide genes are injected into target cells to cause them to self-destruct or kill themselves.

2. Some pathogenic bacteria can withstand changing conditions experienced by their host because genes influencing their antigenicity and other features are highly mutable. Through selection, this hypermutability has allowed some bacteria's genome to develop features that result in high variation in the relevant genes.

3. Three classes of epigenetic inheritance systems are steady-state, structural, and chromatin-marking systems.

4. False. Syntax is unique to humans.

PART EIGHT: BEHAVIORAL GENETICS

- 30 - Concerns Mount over Privacy As Genetic Research Advances

1. Researchers use two common practices to maintain confidentiality of research families' pedigrees: masking, changing the pedigree so that it is obvious to the reader that information is being withheld; and altering, modifying the pedigree by omitting information in such a way that is not obvious to the reader.

2. The community that is being studied in a genetic research project should be involved in the study process to assure that individuals can make truly informed decisions.

3. Populations that are or have been isolated for significant periods have a smaller amount of genetic variability. That is, they are genetically homogenous. Genetically homogenous populations are valuable for the discovery of specific genetic mutations for heritable conditions.

- 31 - The Evolution of Morality

1. Many traditionalists view justice and morality as human inventions that serve to control individuals in our society.

2. Sympathy and empathy are the core of moral behavior as argued by Mr. de Waal.

3. Mr. de Waal says that social animals live in groups for the benefits of group life, such as protection from predators.

- 32 - Making Geneticists Out of Physicians

1. There are not enough geneticists and genetics counselors to handle the growing number of patients who will be interested in, or eligible for, genetic testing, with the explosion of knowledge as a result of the Human Genome Project. Primary care physicians will become increasingly responsible for disseminating this information and for interpreting and explaining the results of genetic testing to their patients.

2. Ethical, Legal, and Social Implications of Human Genome Research.

3. Of the primary care physicians and psychiatrists who responded, fewer than three-quarters of the genetics questions were answered correctly. Professional geneticists in the survey answered nearly 95% of those questions correctly.

- 33 - Mechanisms and Evolution of Aging

1. Some of the strongest candidates of genes for aging and longevity are those that regulate the processes of somatic maintenance and repair.

2. All the "Age" mutations tested so far are resistant to oxidative stress, thermal stress, and UV radiation — with the degree of resistance often proportional to the extension in life-span.

3. The *clk* mutants display a range of phenotypes — including slow development, a slow cell cycle, slow and irregular behavioral rhythms, and extended life-span.